PERSPECTIVES IN MATHEMATICS, Vol. 6

J. Coates and S. Helgason, editors

Analytic Properties of Automorphic L-Functions

Analytic Properties of Automorphic *L*-Functions

Stephen Gelbart

Department of Mathematics
The Weizmann Institute of Science
Rehovot, Israel

Freydoon Shahidi

Department of Mathematics
Purdue University
West Lafayette, Indiana

ACADEMIC PRESS, INC.
Harcourt Brace Jovanovich, Publishers

Boston San Diego New York
Berkeley London Sydney
Tokyo Toronto

ACADEMIC PRESS, INC.
1250 Sixth Avenue, San Diego, CA 92101

United Kingdom Edition published by
ACADEMIC PRESS INC. (LONDON) LTD.
24-28 Oval Road, London NW1 7DX

Work on this volume was partially supported by NSF grant DMS-8521179 and U.S.-Israel Binational Science Foundation Grant #84-10039.

Library of Congress Cataloging-in-Publication Data

Gelbart, Stephen S., Date–
Analytic properties of automorphic L-functions / Stephen Gelbart, Freydoon Shahidi.
p. cm—(Perspectives in mathematics; vol. 6)
Bibliography: p.
Includes index.
ISBN 0-12-279175-4
1. L-functions. 2. Functions, Automorphic. I. Shahidi, Freydoon. II. Title. III. Series.
QA246.G45 1988
512'.73—dc19 88-14598
CIP

88 89 90 91 9 8 7 6 5 4 3 2 1
Printed in the United States of America

TABLE OF CONTENTS

Acknowledgements

The first draft of this manuscript was prepared in the Fall of 1986. We wish to thank R.P. Langlands for his suggestion that a report such as this one be written. His influence on us, and on the field in general, has been immense.

The first named author also wishes to record his indebtedness to I. Piatetski-Shapiro for numerous enlightening discussions about automorphic L-functions, over the period of many years.

The production of this manuscript, from initial stages to camera ready copy, was done by Ruby Musrie of the Weizmann Institute. We are deeply grateful to her for her efficiency, expertise, and good cheer.

INTRODUCTION

The purpose of this monograph is to put into perspective recent work on the automorphic L-functions attached by Langlands to reductive algebraic groups. Whereas Borel's earlier survey on this subject [Bo1] was devoted mainly to related classification and parametrization problems in representation theory, the present work is exclusively concerned with the *analytic* properties of these L-functions. In particular, attention is focused on recent attacks on the general conjecture of Langlands (Conjecture L below) describing the functional equation and meromorphic properties of these L-functions.

To describe these matters in more detail, let us fix the following data:

G –a connected reductive linear algebraic group defined over an $\mathbb{A}$-field F,

$\pi = \otimes \pi_v$– an irreducible unitary automorphic cuspidal representation of the adele group of G,

and

r – a finite dimensional representation of the L-group ${}^LG$ of G.

We refer to such data as the "L-function data" (G, π, r), and we call a finite place v of F *unramified* (for G and π) if:

(1) G is "unramified" over F_v, i.e., as a group over F_v, G is quasi-split and splits over some unramified extension of F_v; and

(2) π_v is "unramified", i.e., the restriction of π_v to a certain maximal compact subgroup of $G_v = G(F_v)$ contains the identity representation. In this case, π_v corresponds via the Satake homomorphism to a ${}^LG^{\circ}$-semisimple conjugacy class t_v in the local L-group ${}^LG_v$, and the *local Langlands L-factor* is defined by

$$L(s, \pi_v, r_v) = \{det[1 - r_v(t_v)q_v^{-s}]\}^{-1} .$$

(Here q_v denotes the residual characteristic of F_v, and $r_v = r \circ \eta_v$ is the corresponding representation of ${}^LG_v$, the L-group of G as a group over F_v, obtained by composing r with the natural homomorphism $\eta_v : {}^L G_v \rightarrow {}^L G$. See I.2.3 and Remark I.2.5.4 for more details.)

Now let S be a finite set of places of F (including the archimedean places) with the property that π_v and G are unramified for each $v \notin S$. The *global* L-function attached to the data (G, π, r) is then given by the Euler product

$$L_S(s, \pi, r) = \prod_{v \notin S} L(s, \pi_v, r_v) \; . \tag{0.1}$$

This function was introduced by Langlands nearly twenty years ago ([La1]). Details and background for the concepts recalled above may be found in [Bo1]; as introductions to [Bo1] (and the present work) we mention [Art1] and [Ge1]. A brief description of the correspondence $\pi_v \longrightarrow t_v$ may be found in Section I.2.3 below.

In [La1] the theory of spherical functions was used to establish the convergence of the Euler product (0.1) in some right half-plane $Re(s) > s_0$ with s_0 depending on the data (G, π, r). This proof exploits only the fact that π is an irreducible unitary (not necessarily automorphic or cuspidal) representation of $G_{\mathbb{A}}$. In case π is actually automorphic cuspidal, Langlands' theory of Eisenstein series (cf. [La2]) is used to further extend $L_S(s, \pi, r)$ from a holomorphic function in $Re(s) > s_0$ to a meromorphic function in $\mathbb{C}$ (at least for a broad class of data (G, π, r) as explained in Section I.2 below).

Shortly after [La1], Langlands conjectured the following:

Conjecture L. (cf. [La3]). Given *any* G, π, r, and S as above, the Euler product $L_S(s, \pi, r)$ continues to a meromorphic function in $\mathbb{C}$ which has only *finitely* many poles and satisfies a standard functional equation relating its value at s to $1 - s$.

Implicit in the statement of this Conjecture is the need to define local "Euler factors" $L(s, \pi_v, r_v)$ at the places v in S as well, and to prove a functional equation for the resulting (completed) Euler product of the form

$$L(s, \pi, r) = \varepsilon(s, \pi, r) L(1 - s, \pi, \tilde{r})$$

with $\varepsilon(s, \pi, r)$ an "exponential" function of s.

Today it is possible to establish this Conjecture for a much wider class of (G, π, r) than was thought possible just a few years ago. Basically two different approaches to the Conjecture have been proven to be successful. The first is Langlands' original approach based on his theory of Eisenstein series ([La1]). As already mentioned, this approach initially yielded only the meromorphic continuation in many special cases. In particular, it gave very little (if any) information on the finiteness of poles or the functional equation. Subsequently, Langlands' approach was refined in [Sha 1-8], with π assumed to be "non-degenerate"; for such representations, the theory of "local coefficients" makes it possible to establish not only a functional equation for $L(s, \pi, r)$, but also the finiteness of poles in a very wide variety of cases ([Sha 1-9]).

A second approach to Langlands' Conjecture was developed first for $G = GL(2)$ by Jacquet and Langlands ([JL]). In this approach, certain integral expressions are introduced which are closely related to $L(s, \pi, r)$; the burden of proof is then shifted to understanding the analytic properties of these "zeta-integrals". Modification of these integrals were subsequently developed for $GL(n)$ by Godement, Jacquet, Piatetski-Shapiro and Shalika in a long series of papers written throughout the 1970's. Finally, in the last few years, these methods have been pushed through for quite general groups, thanks mainly to the efforts of Piatetski-Shapiro (and several others to be named below). Because these zeta-integrals provide an explicit construction of the corresponding L-functions, we shall refer to this approach as the method of "explicit construction of L-functions".

In this survey, we shall describe all the results that are known (to us) concerning the analytic continuation and functional equation of $L_S(s, \pi, r)$. For pedagogical reasons, we treat the two approaches outlined above in parallel, and in chronological order of development.

Advice to the Reader.

We have attempted to keep this work shorter by omitting definitions and computations which are readily accessible in the literature. On the other hand, we have included details of:

(1) known theories which have not previously been described in a form suitable for our purposes (cf. our treatment of the "L-function machine" of Jacquet-Langlands in Section I.1);

(2) computations which are basic to the subject but perhaps too elementary to have been written out previously (cf. our description of the correspondence $\pi_v \longleftrightarrow t_v$ in Section I.2.3); and

(3) new approaches to old theories which – unlike the original approaches – are susceptible to generalization (cf. the new approach to Jacquet's $GL(2) \times GL(2)$ theory in the Appendix to §II.1.2).

The expert in automorphic representations can skip immediately to Chapter III, where the most recent (and partly unpublished) results of Shahidi, and of Piatetski-Shapiro, Rallis, Gelbart, Garrett, and others, are discussed. On the other hand, the reader who is primarily interested in learning how representation theory is used to establish the analytic properties of these L-functions can focus his attention on (the introductory) Chapter I.

CHAPTER I.

FIRST STEPS (1965-1970)

We start with the work of Jacquet and Langlands even though it postdates Langlands' original theory of Euler products by a few years. This work is perhaps easier to appreciate since it treats "only" the group $GL(2)$, and gives an exhaustive treatment of the corresponding "Hecke-type" L-functions $L(s,\pi) = L(s,\pi,r)$, with $r: GL_2(\mathbb{C}) \longrightarrow GL_2(\mathbb{C})$ the "standard" representation. On the other hand, [La1] deals with partial results for more general groups, and is best appreciated only after the proper background, classical and otherwise, has been mastered.

§1. AN ANALYSIS OF THE METHOD OF JACQUET-LANGLANDS

1.1. Cuspidal representations and L-functions for $GL(2)$.

By an *automorphic cuspidal representation* π *of* $GL(2)$ over $\mathbb{Q}$ we understand a collection of irreducible representations $\{\pi_p\}$, one for each prime p of $\mathbb{Q}$ (including the "infinite" prime $p=\infty$), satisfying the following properties:

(i) for each $p \leq \infty$, π_p is an irreducible unitary representation of $GL_2(\mathbb{Q}_p)$ (or $GL_2(\mathbb{R})$ if $p=\infty$), and for almost every $p<\infty$, the restriction of π_p to the maximal compact subgroup $K_p = GL_2(\mathbb{Z}_p)$ contains the trivial representation of K_p, i.e., π_p is class 1 ("spherical") or *unramified*; and

(ii) The (restricted) tensor product representation

$$\pi = \otimes_{p\leq\infty}\pi_p$$

is an irreducible unitary representation of $G_{\mathbb{A}} = \Pi'_p GL_2(\mathbb{Q}_p)$ which occurs discretely in some space of cusp forms on $GL(2)$; more precisely, if

$$Z_{\mathbb{A}} = \left\{ \begin{bmatrix} a & 0 \\ 0 & a \end{bmatrix} \ : \ a \in \mathbb{A}^x \right\}$$

denotes the center of $GL_2(\mathbb{A})$, then π is equivalent to a minimally invariant subspace H_π of the space of χ-central cusp forms $L_0^2(\chi)$, namely those φ on $GL_2(\mathbb{Q})\backslash GL_2(\mathbb{A})$ satisfying

$$\varphi(gz) = \chi(a)\varphi(g) \quad \text{for} \quad z = \begin{bmatrix} a & 0 \\ 0 & a \end{bmatrix} \quad \text{in} \quad Z_{\mathbb{A}} \ ,$$

where χ is a unitary character of $Q^x \backslash \mathbb{A}^x$,

$$\int_{Z_{\mathbb{A}} GL_2(\mathbb{Q})\backslash GL_2(\mathbb{A})} |\varphi(g)|^2 dg < \infty$$

and

$$\int \varphi\left(\begin{bmatrix} 1 & x \\ 0 & 1 \end{bmatrix} g \right) dx = 0 \quad \text{for almost all } g \text{ in } GL_2(\mathbb{A}) \ .$$

(This last condition is the condition of *cuspidality.*)

For background material on these notions, and connections with the classical theory of automorphic forms, the reader is referred to [Ge2]. For a detailed treatment of the notion of an automorphic representation over an arbitrary $\mathbb{A}$-field F – even in the context of a general reductive group G, the reader is referred to [BoJa]. The point is that for $GL(2)$ over $\mathbb{Q}$, the notion of an automorphic cuspidal representation $\pi = \otimes \pi_p$ is the natural generalization of the classical notion of a cusp form

$$f(z) = \sum_{n=1}^{\infty} a_n e^{2\pi i n z}$$

on $SL_2(\mathbb{Z})$, and the work of Jacquet-Langlands amounts to a representation theoretic treatment of the L-function

$$\text{(I.1.1.1)} \qquad L(s,f) = (2\pi)^{-s}\Gamma(s)\left(\sum_{n=1}^{\infty} \frac{a_n}{n^s} \right) = \int_0^\infty f(iy) y^{s-1} dy$$

attached to f (by Hecke).

Let us suppose, as usual, that f is an eigenfunction for all the Hecke operators $T(p)$, with $a_1 = 1$. Thus

$$L(s,f) = (2\pi)^{-s}\Gamma(s)\prod(1 - a_p p^{-s} + p^{k-1-2s})^{-1}$$

with k the weight of f. In order to interpret the factor $(1 - a_p p^{-s} + p^{k-1-2s})^{-1}$ as a local (Langlands) L-factor $L(s,\pi_p,r)$, we need to recall a few basic facts from the dictionary

$$f \longrightarrow \pi_f = \otimes \pi_p$$

relating classical cusp forms to automorphic cuspidal representations of $GL(2)$ (cf. Chapters 4 and 5 of [Ge2]). In particular, we recall that any unramified representation π_p of $GL_2(\mathbb{Q}_p)$ is of the form

$$\pi_p = Ind_{B_p}^{GL_2(\mathbb{Q}_p)} \mu_1\mu_2$$

where μ_1, μ_2 are (unramified) characters of $\mathbb{Q}_p^x$ of the form $|x|_p^{s_i}$, s_i in $\mathbb{C}$, and the pair (μ_1, μ_2) defines a character of the Borel subgroup B_p through the formula

$$\mu_1\mu_2\left(\begin{bmatrix} a_1 & x \\ 0 & a_2 \end{bmatrix}\right) = \mu_1(a_1)\mu_2(a_2) .$$

Furthermore, if such a π_p happens to be a local component of the cuspidal representation π_f attached to f, then the characters μ_1, μ_2 are explicitly determined by the Fourier coefficients of f through the formulas

$$a_p = p^{(k-1)/2}(\mu_1(p) + \mu_2(p))$$

and

$$\mu_1\mu_2 \equiv 1 .$$

For $G = GL(2)$, it happens that the L-group ${}^LG = GL_2(\mathbb{C})$, and the conjugacy class t_p *) in ${}^LG$ (canonically associated to the unramified

---

*) We give a general explanation of the correspondence $\pi_p \longrightarrow t_p$ in Section 2.3 below; appropriate references are [Bo1] and [Car].

representation $\pi_p = Ind\ \mu_1\mu_2)$ is

$$t_p = \begin{bmatrix} \mu_1(p) & 0 \\ 0 & \mu_2(p) \end{bmatrix}.$$

Thus, letting r denote the "standard" (two-dimensional) representation of ${}^LG = GL_2(\mathbb{C})$ (taking each g in $GL_2(\mathbb{C})$ to itself), it is clear that

$$\begin{aligned} L(s,\pi_p,r) &= \{det[I - r(t_p)p^{-s}]\}^{-1} \\ &= (1 - a_p p^{-s'} + p^{k-1-2s'})^{-1} \end{aligned}$$

where $s' = \frac{k-1}{2} + s$, i.e.,

$$\prod_{p<\infty} L(s,\pi_p,r) = \sum_{n=1}^{\infty} \frac{a_n}{n^{s'}}.$$

1.2. Global zeta-integrals and their factorization.

Having recalled how the relation between $\{\pi_p\}$ and the automorphic L-function $L(s,\pi_f)$ $(= L(s',f))$ describes the setting for [JL], it remains to explain how the analytic properties of $L(s,\pi_f)$ are derived from those of $\pi = \otimes\pi_p$ via "zeta-integrals".

The idea is to first introduce a *global zeta-integral* of the form

(I.1.2.1) $$\mathcal{Z}(\varphi,s) = \int_{\mathbb{Q}^x\backslash\mathbb{A}^x} \varphi\left(\begin{bmatrix} a & 0 \\ 0 & 1 \end{bmatrix}\right) |a|_{\mathbb{A}}^{s-1/2} d^x a$$

where φ is any function in the subspace H_π of $L_0^2(\chi)$ realizing π. (Note how this integral generalizes the classical Mellin transform which appears in (I.1.1.1).) If φ is suitably nice, but still arbitrary in a dense subspace*)

*) For example, φ might be any function in H_π which is *right K-finite* for the standard maximal compact subgroup $K_{\mathbb{A}} = \Pi_p K_p$, i.e., the right translates of φ by k in $K_{\mathbb{A}}$ span a finite dimensional space of functions; cf. Section (3.1) of [God].

of H_π, then φ is *rapidly decreasing* at infinity. This means that for any given compact set C in $G_{\mathbb{A}}$, and $t \geq 0$, we have (for *any* N)

$$\varphi\left(\begin{bmatrix} a & 0 \\ 0 & 1 \end{bmatrix} g\right) \leq C_N |a|^{-N} \tag{I.1.2.2}$$

for g in C and $|a| > t$. From this it follows that the integral defining $\mathcal{Z}(\varphi, s)$ converges for all values of s. Indeed, from the fact that φ is χ-central, and invariant on the left by $w = \begin{bmatrix} 0 & 1 \\ -1 & 0 \end{bmatrix}$, which is in $G_{\mathbb{Q}}$, we have

$$\begin{aligned} \varphi\left(\begin{bmatrix} a^{-1} & 0 \\ 0 & 1 \end{bmatrix}\right) &= \chi^{-1}(a)\varphi\left(\begin{bmatrix} 1 & 0 \\ 0 & a \end{bmatrix}\right) = \chi^{-1}(a)\varphi\left(w^{-1}\begin{bmatrix} a & 0 \\ 0 & 1 \end{bmatrix} w\right) \\ &= \chi^{-1}(a)\varphi\left(\begin{bmatrix} a & 0 \\ 0 & 1 \end{bmatrix} w\right) . \end{aligned}$$

Thus $\varphi\left(\begin{bmatrix} a & 0 \\ 0 & 1 \end{bmatrix}\right)$ is also rapidly decreasing "at zero" (since the right translate of φ by w is still rapidly decreasing at infinity). Hence $\mathcal{Z}(\varphi, s)$ defines an *entire* function in $\mathbb{C}$ (and satisfies a simple functional equation which we shall soon recall).

The problem now is to relate these nice analytic functions $\mathcal{Z}(\varphi, s)$ to our automorphic L-function $L(s, \pi)$. The first step is to write $\mathcal{Z}(\varphi, s)$ as a product of *local zeta-integrals* by exploiting the Fourier expansion

$$\varphi(g) = \sum_{\xi \in \mathbb{Q}^x} W_\varphi^\psi\left(\begin{bmatrix} \xi & 0 \\ 0 & 1 \end{bmatrix} g\right) . \tag{I.1.2.3}$$

Here ψ is a fixed non-trivial character of $\mathbb{Q}\backslash\mathbb{A}$, and W_φ is the ψ-th Fourier coefficient of φ, namely

$$W_\varphi^\psi(g) = \int_{\mathbb{Q}\backslash\mathbb{A}} \varphi\left(\begin{bmatrix} 1 & x \\ 0 & 1 \end{bmatrix} g\right) \overline{\psi(x)} dx . \tag{I.1.2.4}$$

We recall (cf. [Ge2], Chapter 3) that (I.1.2.3) merely generalizes the familiar Fourier expansion of a classical cusp form. We also note that

$W_\varphi(g)$ is rapidly decreasing "at infinity", since φ satisfies (I.1.2.2). Plugging (I.1.2.3) into (I.1.2.1), and collapsing the double integration into one, yields the basic identity

$$\mathcal{Z}(\varphi, s) = \int_{\mathbb{A}^x} W_\varphi\left(\begin{bmatrix} a & 0 \\ 0 & 1 \end{bmatrix}\right) |a|^{s-1/2} d^x a \ , \tag{I.1.2.5}$$

valid for $Re(s)$ sufficiently large.

In order to factor the right hand side of this identity as an "Euler product" of local integrals, we must recall some special properties of the functions W_φ. For each φ as above, $W_\varphi(g)$ is not only rapidly decreasing at infinity, but also satisfies the condition

$$W\left(\begin{bmatrix} 1 & x \\ 0 & 1 \end{bmatrix} g\right) = \psi(x) W(g) \ . \tag{I.1.2.6}$$

Thus the map $\varphi \mapsto W_\varphi$ intertwines a dense subspace of H_π with the so-called *ψ-Whittaker model for* π, i.e., the space of "smooth", right K-finite, χ-central, rapidly decreasing functions on $G_{\mathbb{A}}$ satisfying (I.1.2.6), and realizing π under the natural right action of $G_{\mathbb{A}}$ (or rather the *Hecke algebra* of $G_{\mathbb{A}}$; cf. §3.4 of [God] for details). It is well known ([JL], §9) that such a model is unique, and equals the tensor product of local such Whittaker models $\mathcal{W}(\pi_p, \psi_p)$, where $\pi = \otimes \pi_p$ and $\psi = \Pi \psi_p$. In particular, we may assume W_φ is of the form

$$W_\varphi(g) = \prod_p W_p(g) \ , \tag{I.1.2.7}$$

where each $W_p \in \mathcal{W}(\pi_p, \psi_p)$, and for almost all p, $W_p(k) \equiv 1$ for $k \in K_p = GL_2(\mathbb{Z}_p)$. (The p-adic analogue of smoothness is now that W_p is locally constant.) As we shall soon see, these W_p are such that (1) the *local zeta-integrals*

$$\mathcal{Z}(W_p, s) = \int_{\mathbb{Q}_p^x} W_p\left(\begin{bmatrix} a & 0 \\ 0 & 1 \end{bmatrix}\right) |a|^{s-1/2} d^x a \tag{I.1.2.8}$$

converge absolutely for $Re(s) > s_0$, and (2) their product (over all p) also converges absolutely for $Re(s) > s_0$; see Remark I.1.4.4 below. Thus we obtain the Euler product expansion

$$\mathcal{Z}(\varphi, s) = \prod_{p \leq \infty} \mathcal{Z}(W_p, s) \tag{I.1.2.9}$$

valid for $Re(s) \gg 0$.

1.3. Local zeta-integrals.

Having reduced matters to the local theory (thanks to the Euler product expansion (I.1.2.9)), it remains to analyze the local zeta integrals $\mathcal{Z}(W_p, s)$ and their relation with the Euler factors $L(s, \pi_p)$ defined by Jacquet and Langlands. The reader should keep in mind that this is exactly the same game plan which was exploited first in Tate's thesis [Tate1]. (There the Euler factor was the p-th component $(1-\chi(p)p^{-s})^{-1}$ of Dirichlet's L-function $L(s,\chi)$, and the local zeta-integral was of the form $\mathcal{Z}(f,s) = \int_{\mathbb{Q}_p^x} f(x)|x|^s d^x x$, with f a Schwartz-Bruhat function on $\mathbb{Q}_p$.)

In the present context, our first step is to analyze the analytic properties of the *local* zeta-integrals $\mathcal{Z}(W,s)$. We begin by recalling the asymptotic behavior of Whittaker functions at 0 as well as at infinity (at least for *non*-archimedean $\mathbb{Q}_p$). It is a simple fact that for any W in $\mathcal{W}(\pi_p)$,

$$W\begin{pmatrix} a & 0 \\ 0 & 1 \end{pmatrix} = 0 \quad \text{for} \quad |a| \quad \text{large}\ . \tag{I.1.3.1}$$

On the other hand, for small $|a|$, there exists a finite set of "finite" functions $c_i(a)$ (depending only on π_p) such that

$$W\begin{pmatrix} a & 0 \\ 0 & 1 \end{pmatrix} \sim \sum_{i=1}^{N} \alpha_i c_i(a)\ , |a| \ \text{small}\ . \tag{I.1.3.2}$$

Here "finite" means that the space of right translates by $\mathbb{Q}_p^x$ spans a finite-dimensional space; finite functions on Q_p^x are thus characters, integer powers of the valuation function $v(a)$, or products and linear combinations thereof.

The identity (I.1.3.1) is indeed elementary since the functional equation (I.1.2.6), together with local constancy, implies that for each small $|x|$,

(I.1.3.3)

$$W\left(\begin{bmatrix} a & 0 \\ 0 & 1 \end{bmatrix}\right) = W\left(\begin{bmatrix} a & 0 \\ 0 & 1 \end{bmatrix}\begin{bmatrix} 1 & x \\ 0 & 1 \end{bmatrix}\right) = W\left(\begin{bmatrix} 1 & ax \\ 0 & 1 \end{bmatrix}\begin{bmatrix} a & 0 \\ 0 & 1 \end{bmatrix}\right)$$
$$= \psi(ax) W\begin{pmatrix} a & 0 \\ 0 & 1 \end{pmatrix};$$

in particular, if we pick $|a|$ so large that $\psi(ax) \neq 1$, then $W\begin{pmatrix} a & 0 \\ 0 & 1 \end{pmatrix} = 0$. The asymptotic expansion *at 0*, however, lies deeper. In [JL], this expansion is established by a case by case analysis of representation types for π_p (principal series vs. special representations vs. "supercuspidals"). Nowadays, we also know how to derive it from more general results of Casselman and Shalika (more about this in §III.1.5).

As an immediate consequence of these asymptotic expansions we obtain the following results for our local zeta-integrals $\mathcal{Z}(W,s)$:

(i) The integral defining $\mathcal{Z}(W,s)$ converges for $Re(s)$ greater than some s_0 (independent of W);

(ii) The resulting functions $\mathcal{Z}(W,s)$ are in fact rational functions in p^{-s}, hence are defined as meromorphic functions *in all of* $\mathbb{C}$;

(iii) The family of rational functions $\{\mathcal{Z}(W,s)\}$ admits a common denominator (as W runs through $W(\pi_p,\psi)$), i.e., a polynomial P such that $P(p^{-s})\mathcal{Z}(W,s)$ belongs to $\mathbb{C}[p^{-s},p^s]$ for all W in $\mathcal{W}(\pi_p)$.

A closely related result is the following:

(iv) There exists a W^* in $\mathcal{W}(\pi_p,\psi_p)$ with the property that $\mathcal{Z}(W^*,s) \equiv 1$.

1.4. More local theory.

The key missing ingredients in the local theory are now the "functional equation" and the "unramified analysis". The functional equation asserts that there exists a meromorphic function $\gamma(\pi_p, \psi_p, s)$ (which is in fact rational in p^{-s} for $p < \infty$) such that

(I.1.4.1) $$\tilde{\mathcal{Z}}(W^w, 1-s) = \gamma(\pi_p, \psi_p, s)\mathcal{Z}(W, s) .$$

Here $W^w(g) = W(gw)$, and $\tilde{\mathcal{Z}}(W,s) = \int_{\mathbb{Q}_p^x} W\left(\begin{pmatrix} a & 0 \\ 0 & 1 \end{pmatrix}\right) |a|^{s-1/2} \cdot$ $\cdot\chi^{-1}(a)d^x a$, with χ_p the central character of π_p.Taking for granted the fact that these integrals $\mathcal{Z}(W,s)$ and $\tilde{\mathcal{Z}}(W^w, 1-s)$ are rational functions in p^{-s} (or at least meromorphic functions), this functional equation reduces to the assertion that – for almost every s – there is a "unique" functional ℓ_s on the space of (Kirillov) functions $\{\xi(x) = W\begin{pmatrix} x & 0 \\ 0 & 1 \end{pmatrix} : W \in \mathcal{W}(\pi, \psi)\}$ such that $\ell_s(\xi_{\pi\left(\begin{smallmatrix} a & 0 \\ 0 & 1 \end{smallmatrix}\right)W}) = |a|^{1/2-s}\ell_s(\xi_W)$; indeed $\mathcal{Z}$ and $\tilde{\mathcal{Z}}$ each define such a functional, so $\gamma(\pi_p, \psi_p, s)$ is simply the resulting constant of proportionality.

Concerning the unramified analysis, we recall that π_p *unramified* means that π_p is of the form $Ind\ \mu_1\mu_2$ with μ_1, μ_2 unramified characters of $\mathbb{Q}_p^x$. We also suppose that W^0 denotes the (unique) right K_p-invariant function in $\mathcal{W}(\pi_p, \psi_p)$ such that $W(k) \equiv 1$ for all k in $K_p = GL_2(\mathbb{Z}_p)$. (We are supposing here that ψ_p is also unramified, i.e. $\psi_p(x)$ is identically 1 on $\mathbb{Z}_p$ but not on $p^{-1}\mathbb{Z}_p$.) Then the claim is that

(I.1.4.2) $$\mathcal{Z}(W^0, s) = det[I - r(t_p)p^{-s}]^{-1} = L(s, \pi_p, r_p)$$

with $t_p = \begin{bmatrix} \mu_1(p) & 0 \\ 0 & \mu_2(p) \end{bmatrix}$ the conjugacy class in $GL_2(\mathbb{C})$ canonically associated with π_p. Using a point of view not yet fully understood at the time of [JL], we shall give a slick proof of (I.1.4.2) which has the advantage of being generalizable to the context of arbitrary reductive groups.

In the present unramified context, equation (I.1.3.3) implies that $W^0\left(\begin{bmatrix} a & 0 \\ 0 & 1 \end{bmatrix}\right) = W\left(\begin{bmatrix} p^n & 0 \\ 0 & 1 \end{bmatrix}\right) = 0$ unless $|a| = |p^n u| = p^{-n} \leq 1$. Thus

$$\begin{aligned} \mathcal{Z}^0(W,s) &= \int_{|a|\leq 1} W^0\left(\begin{bmatrix} a & 0 \\ 0 & 1 \end{bmatrix}\right) |a|^{s-1/2} d^x a \\ &= \sum_{n=0}^{\infty} W^0\left(\begin{bmatrix} p^n & 0 \\ 0 & 1 \end{bmatrix}\right) (p^{-n})^{s-1/2} . \end{aligned}$$

Now it is a result of [Shin1] (and a special case of a still more recent general result of [CS] and [Ka] that the unramified Whittaker function for GL_r may be expressed in terms of characters of finite dimensional representations of $GL_r(\mathbb{C})$. In particular, if we let χ_n denote the character of the unique irreducible representation $\rho_{n,0}$ of $GL_2(\mathbb{C})$ whose highest weight is $(n,0)$, $n \geq 0$, then

$$W^0_{\pi_p}\left(\begin{bmatrix} p^n & 0 \\ 0 & 1 \end{bmatrix}\right) = |p^n|^{1/2}\chi_n(t_p) .$$

Thus we have

$$\mathcal{Z}(W^0, s) = \sum_{n=0}^{\infty} \chi_n(t_p)X^n ,$$

with $X = p^{-s}$. It remains to observe that

(I.1.4.3) $$\sum \chi_n(t_p)\chi^n = det[I - t_pX]^{-1} .$$

Indeed, the right hand side of (I.1.4.3) may be rewritten as $\sum_{\ell=0}^{\infty} Sym^\ell(t_p)X^\ell$, where $Sym^\ell(t_p)$ denotes the $\ell - th$ symmetric power of the matrix $t_p = \begin{bmatrix} \mu_1(p) & 0 \\ 0 & \mu_2(p) \end{bmatrix}$, and $\rho_{n,0}(t_p) = Sym^n(t_p)$.

Remark 1.4.4. Let us rewrite (1.4.2) in the form

$$\mathcal{Z}(W^0_p, s) = [(1 - \mu_1(p)p^{-s})(1 - \mu_2(p)p^{-s})]^{-1} .$$

Keep in mind that the integral defining $\mathcal{Z}(W^0_p, s)$ – namely $\int W^0_p\begin{pmatrix} a & 0 \\ 0 & 1 \end{pmatrix}|a|^{s-1/2}d^x a$ – converges for $Re(s) > s_0$ (independent of p). The theory of p-adic spherical functions tells that if *ind* $\mu_1\mu_2$ is unitary *and* class 1, then there exists $a > 0$ (and independent of p) such that $|\mu_i(p)| \leq p^a$. From this it follows that the infinite product $\prod_{p\leq\infty} \mathcal{Z}(W_p, s)$ converges absolutely for $Re(s)$ sufficiently large. For $G = GL(2)$ we also know how to obtain the sharper estimate $|\mu_i(p)| \leq p^{1/2}$ directly from the (known) classification of unitary irreducible representations of G. For general G, such a classification of unitary spherical representations – and consequently such a sharp estimate for $|\mu_i(p)|$ – is not yet available. Thus the theory of spherical functions must be used (as explained in [Bo]).

1.5. Global results for $L_S(s,\pi)$.

We explain now how the local and global theory just outlined yields the required results for $L(s,\pi) = \prod_p L(s,\pi_p)$. Actually, since we haven't yet described how $L(s,\pi_p)$ is defined for "ramified p", we must first restrict our attention to the partial L-function

$$L_S(s,\pi,r) = \prod_{\substack{p \\ \text{unramified}}} L(s,\pi_p,r_p)\ . \tag{I.1.5.1}$$

For each "ramified p", let us pick W_p in $\mathcal{W}(\pi_p)$ (as in 1.3(iv)) such that $\mathcal{Z}(W_p,s) \equiv 1$. (For $p=\infty$, we can at least pick W_p so that $\mathcal{Z}(W_p,s)$ is a non-vanishing entire function of s.) Then for $W = \Pi W_p$, with $W_p = W_p^0$ for each unramified p, we have

$$\mathcal{Z}(\varphi,s) = e^{g(s)} L_S(s,\pi)$$

for some holomorphic function $g(s)$. Thus the holomorphy of $L_S(s,\pi)$ on $\mathbb{C}$ follows immediately from the known holomorphy of the zeta-integral $\mathcal{Z}(\varphi,s)$.

In order to describe the functional equation of $L_S(s,\pi)$ we need first to recall the functional equation for $\mathcal{Z}(\varphi,s)$. But this is easy: since $\varphi(w^{-1}g) = \varphi(g)$, $w = \begin{bmatrix} 0 & 1 \\ -1 & 0 \end{bmatrix}$, we have

$$\begin{aligned}
\mathcal{Z}(\varphi,s) &= \int_{\mathbb{Q}\backslash A^x} \varphi\left(\begin{bmatrix} a & 0 \\ 0 & 1 \end{bmatrix}\right) |a|^{s-1/2} d^x a \\
&= \int \varphi\left(\begin{bmatrix} a^{-1} & 0 \\ 0 & 1 \end{bmatrix}\right) |a|^{-s+1/2} d^x a \\
&= \int \chi(a^{-1})\varphi\left(w^{-1}\begin{bmatrix} a & 0 \\ 0 & 1 \end{bmatrix} w\right) |a|^{(1-s)-1/2} d^x a \\
&= \tilde{\mathcal{Z}}(\varphi^w, 1-s)\ ,
\end{aligned} \tag{I.1.5.2}$$

where $\varphi^w(g) = \varphi(gw)$, and $\tilde{\mathcal{Z}}(\varphi,s) = \int \varphi\left(\begin{bmatrix} a & 0 \\ 0 & 1 \end{bmatrix}\right) \chi^{-1}(a)|a|^{s-1/2} d^x a$. In order to derive from this the functional equation for $L_S(s,\pi)$ we compute (using the local functional equation):

$$\begin{aligned}\mathcal{Z}(\varphi,s) &= \prod_{p\le\infty} \mathcal{Z}(W_p,s)\\ &= \left(\prod_{p\in S} \mathcal{Z}(W_p,s)\right) L_S(s,\pi)\\ &= \left(\prod_{p\in S} \gamma^{-1}(\pi_p,\psi_p,s)\right) \prod_{p\in S} \tilde{\mathcal{Z}}(W_p^w,1-s) L_S(s,\pi)\ .\end{aligned}$$

On the other hand, by the global functional equation,

$$\begin{aligned}\mathcal{Z}(\varphi,s) = \tilde{\mathcal{Z}}(\varphi^w,1-s) &= \prod \tilde{\mathcal{Z}}(W_p^w,1-s)\\ &= \prod_{p\in S} \tilde{\mathcal{Z}}(W_p^w,1-s) L_S(1-s,\pi,\tilde{r})\end{aligned}$$

since for $W_p = W_p^0$,

$$\begin{aligned}\tilde{\mathcal{Z}}(W_p^w,s) &= \left[(1-\mu_1^{-1}(p)p^{-s})(1-\mu_2^{-1}(p)p^{-s})\right]^{-1}\\ &= det[I - r(\tilde{t}_p)p^{-s}]^{-1}\end{aligned}$$

with $\tilde{t}_p = \begin{bmatrix} \mu_1^{-1}(p) & 0 \\ 0 & \mu_2^{-1}(p) \end{bmatrix}$ the conjugacy class in $GL_2(\mathbb{C})$ associated to the "contragredient" representation $\tilde{\pi} = ind\ \mu_1^{-1}\mu_2^{-1}$. Equating these two expressions for $\mathcal{Z}(\varphi,s)$, and choosing W_p so that $\tilde{\mathcal{Z}}(W_p^w,1-s)$ is non-vanishing for $p \in S$, yields the desired functional equation

$$L_S(s,\pi,r) = \left(\prod_{p\in S} \gamma(\pi_p,\psi_p,s)\right) L_S(1-s,\pi,\tilde{r})\ . \tag{I.1.5.3}$$

1.6. Global results for $L(s,\pi)$.

In order to put the functional equation (I.1.5.3) in a better form, it remains to analyze the factors $\gamma(\pi_p,\psi_p,s)$ for the bad primes p; in particular, we must relate the factors $L(s,\pi_p)$ and $\mathcal{Z}(W_p,s)$ for these primes.

We begin by explaining how $L(s,\pi_p)$ may be defined as the "greatest common divisor" of the family of zeta-integrals $\mathcal{Z}(W_p,s)$. Recall from I.1.3(iii) that there exists a polynomial P such that $P(p^{-s})\mathcal{Z}(W,s)$ belongs to $\mathbb{C}[p^{-s},p^s]$ for all W in $\mathcal{W}(\pi_p)$. By a "fractional ideal" in the ring $\mathbb{C}[p^s,p^{-s}]$ we understand a subset $\mathcal{A}$ of $\mathbb{C}(p^{-s})$ (the field of fractions of $\mathbb{C}[p^s,p^{-s}]$) which is a $\mathbb{C}[p^s,p^{-s}]$ module and such that $R\mathcal{A}\subset \mathbb{C}[p^s,p^{-s}]$ for some R in $\mathbb{C}[p^{-s}]$. Note that each fractional ideal is a principal ideal, since $\mathbb{C}[p^s,p^{-s}]$ is the localization of the principal ideal ring $\mathbb{C}[p^s]$.

Now let $\mathcal{A}_p$ denote the $\mathbb{C}[p^s,p^{-s}]$-module generated by all the zeta-integrals $\mathcal{Z}(W_p,s)$ as W_p runs through $\mathcal{W}(\pi_p,\psi_p)$. By 1.3(ii), $\mathcal{A}_p$ is a subspace of $\mathbb{C}(p^{-s})$, and by 1.3(iii), there exists a "common denominator" P such that $P\mathcal{A}_p\subset \mathbb{C}[p^s,p^{-s}]$. Thus $\mathcal{A}_p$ is a fractional ideal of $\mathbb{C}[p^s,p^{-s}]$ in the above sense, and we can let $L(s,\pi_p)$ denote any one of its generators.

By definition, $L(s,\pi_p)$ may be taken to be the quotient of two polynomials R and Q (in p^{-s}) which are relatively prime and such that the denominator has constant term 1. By by 1.3(iv) we know that $\mathcal{Z}(W,s)$ is identically 1 for at least one W_p. Thus the numerator can not have any zeros, and we conclude that $L(s,\pi_p)$ may be defined uniquely by the condition that

$$L(s,\pi_p)=Q(p^{-s})^{-1}$$

with $Q(0)=1$. Note then $\mathcal{A}_p = L(s,\pi_p)\,\mathbb{C}[p^s,p^{-s}]$. For unramified π_p, we have $L(s,\pi_p)=\mathcal{Z}(s,W_p^0)=det[I-r(t_p)p^{-s}]^{-1}$.

We note that $L(s,\pi_p)$, as just defined, really is the greatest common divisor of the family of zeta-integrals $\mathcal{Z}(W_p,s)$. Indeed, $L(s,\pi_p)$ "divides" each $\mathcal{Z}(W_p,s)$ in the sense that $\frac{\mathcal{Z}(W_p,s)}{L(s,\pi_p)}$ is always entire. On the other hand, since $\mathcal{A}_p=L(\pi_p,s)\,\mathbb{C}[p^{-s},p^{-s}]$ for any π_p, and $1\in\mathbb{C}[p^{-s},p^s]$, it follows that $L(s,\pi_p)$ must equal $\mathcal{Z}(W_p,s)$ for some *appropriately* chosen W_p in $\mathcal{W}(\pi_p,\psi_p)$. Thus $L(s,\pi_p)$ is indeed the *greatest* such divisor, i.e., if $L'(s,\pi_p)$ also divides all $\mathcal{Z}(W_p,s)$, then it must divide $L(s,\pi_p)$ as well.

We may similarly define

$$L(s,\tilde{\pi}_p) = \frac{1}{\tilde{Q}(p^{-s})}$$

as the g.c.d. of the "contragredient" zeta-integrals $\tilde{\mathcal{Z}}(W_p, s)$. (The term "contragredient" is apt since the functions $W_p(g)\chi_p^{-1}(det\ g)$ generate the Whittaker model of $\mathcal{W}(\tilde{\pi}_p, \psi_p)$.)

Let us turn finally to the functional equation for the completed L-function

$$L(s,\pi) = \prod_p L(s,\pi_p)\ .$$

Choosing W_p so that $\mathcal{Z}(W_p, s) = L(s,\pi_p)$, we have

$$\begin{aligned}
L(s,\pi) &= \prod \mathcal{Z}(W_p, s) = \mathcal{Z}(\varphi, s) \\
&= \tilde{\mathcal{Z}}(\varphi^w, 1-s) = \left(\prod_{p\in S} \tilde{\mathcal{Z}}(W^w, 1-s)\right) \prod_{p\notin S} L(1-s,\tilde{\pi}_p) \\
&= \left(\prod_{p\in S} \frac{\tilde{\mathcal{Z}}(W^w, 1-s)}{L(1-s,\tilde{\pi}_p)}\right) L(1-s,\tilde{\pi}) \\
&= \left(\prod_{p\in S} \frac{\gamma(\pi_p,\psi_p,s)\mathcal{Z}(W,s)}{L(1-s,\tilde{\pi}_p)}\right) L(1-s,\tilde{\pi}) \\
&= \left(\prod_{p\in S} \varepsilon(\pi_p,\psi_p,s)\right) L(1-s,\tilde{\pi})
\end{aligned}$$

where

$$\varepsilon(\pi_p,\psi_p,s) = \frac{\gamma(\pi_p,\psi_p,s)L(s,\pi_p)}{L(1-s,\tilde{\pi}_p)}\ .$$

Here we used both the global and local functional equations for the zeta-integral.

Theorem. The automorphic L-function $L(s,\pi) = \prod L(s,\pi_p)$, initially defined only in some right half-plane, extends to an entire function of s in $\mathbb{C}$, and satisfies the functional equation

$$L(s,\pi) = \varepsilon(s,\pi)L(1-s,\tilde{\pi}) \qquad \text{(I.1.6.1)}$$

with $\varepsilon(s,\pi) = \prod_{p\in S} \varepsilon(s,\pi_p,\psi)$ a "monomial" function of s.

Proof. We need only check that $\varepsilon(s,\pi)$ is "monomial", i.e., a function without zeros or poles. Since we may choose W_p so that $\frac{\mathcal{Z}(W_p,s)}{L(s,\pi_p)} = 1$, it is clear from the local functional equation

$$\frac{\mathcal{Z}(W_p^w,1-s)}{L(1-s,\tilde{\pi}_p)} = \varepsilon(s,\pi_p,\psi_p)\frac{\mathcal{Z}(W_p,s)}{L(s,\pi_p)}$$

(and the definition of $L(s,\tilde{\pi}_p)$) that $\varepsilon(s,\pi_p,\psi_p)$ is entire. On the other hand, applying this functional equation twice gives

$$\varepsilon(s,\pi_p,\psi_p)\varepsilon(1-s,\tilde{\pi}_p,\psi_p) = \chi_\pi(-1)\ ,$$

and this implies $\varepsilon(s,\pi_p,\psi)$ has no zeros.

Note that (I.1.6.1) is indeed a "better form" of the functional equation (I.1.5.3). First of all, it involves an L-function taken over *all* the places of $\mathbb{Q}$. Secondly, it involves an *exponential* factor of proportionality $\varepsilon(s,\pi)$; thus the zeros and (non-existent) poles of $L(s,\pi)$ are indeed symmetric across the line of symmetry $Re(s) = \frac{1}{2}$.

1.7. Description of the L-function machine.

For future reference, let us sum up the steps taken to describe the analytic continuation and functional equation of $L(s,\pi) = L(s,\pi,r)$.

Step 1. Establish a "Basic Identity" for the global zeta-integral, plus an Euler product expansion.

Step 2. Analyze the meromorphic behavior of the global zeta-integrals, *and* their functional equation.

Step 3. Prove a Functional Equation and Meromorphic Continuation for the *local* zeta-integrals.

Step 4. Equate the *unramified* local zeta-integrals with the "Langlands factors"

$$L(s,\pi_p) = \{det[I - r(p)p^{-s}]\}^{-1}\ .$$

Step 5. Establish basic properties of the local zeta-integrals, namely existence of common denominators and definition and properties of the g.c.d. $L(s, \pi_p)$.

We note that in Tate's theory of L-functions for $GL(1)$, Steps 1 and 4 are immediate, whereas Steps 3 and 5 produce the well-known abelian factors $\gamma(s, \chi_p, \psi_p)$ and $\varepsilon(s, \chi_p, \psi_p)$. For $GL(2)$ and r the standard representation, these steps at least now seem natural to us, thanks to the work of Jacquet-Langlands. However, for more general representations r, or more general groups G, any one (or part) of these steps can present non-trivial obstacles.

Henceforth, we shall refer to the five step procedure just outlined as the (local-global) "L-function machine".

§2. EISENSTEIN SERIES & LANGLANDS' "EULER PRODUCTS"

In [La1], a general theory of automorphic L-functions is broached for the first time. Given datum (G, π, r) (arbitrary except for the technical assumption that G is a split, semisimple algebraic group of adjoint type), the theory of spherical functions is used to prove that there exists a real number s_0, depending only on G, such that Euler products of the form

$$L_S(s, \pi, r) = \prod [det(I - r_v(t_v)q^{-s})]^{-1}$$

converge in some right half-plane $Re(s) > s_0$. (Cf. Remark I.1.4.4.) For a detailed discussion of the proof of this fact in case G is an arbitrary connected reductive group and π is any irreducible (not necessarily automorphic) unitarizable representation of $G_{\mathbb{A}}$, see [La3] and Section 13 of [Bo1]. As we shall see in Chapter III below, sharper uniform estimates for the line of convergence follow from [Shahidi 8] at least when π is *generic* automorphic cuspidal.

Our purpose in this section is to explain how Langlands' theory of Eisenstein series establishes the meromorphic continuation of $L_S(s, \pi, r)$ for at least one non-trivial r of ${}^L G$ (and all but three of the simple groups G). Before getting lost in the notation required for the general theory, let us explain the main ideas by way of the simplest possible example, namely Dirichlet's L-functions.

2.1. The example of $L(s, \chi)$.

In order to deal with the functions $L(s, \chi)$, χ a character of $\mathbb{Q}^x \backslash \mathbb{A}^x$, we need to consider Eisenstein series on the group $H = SL(2)$; these are defined by the equation

$$E(s, f, h) = \sum_{\gamma \in B_f \backslash H_F} f_s(\gamma h) \ , \qquad \text{(I.2.1.1)}$$

where B is the Borel subgroup $\left\{\begin{bmatrix} a & x \\ 0 & a^{-1} \end{bmatrix}\right\}$, and f_s belongs to the space of the induced representation

$$Ind_{B_{\mathbb{A}}}^{H_{\mathbb{A}}} \chi(a)|a|_{\mathbb{A}}^s = I(s,\chi) \ . \tag{I.2.1.2}$$

The theory of Eisenstein series tells us that the series defining $E(s,f,h)$ converges absolutely for $Re(s) \gg 0$, and that the resulting holomorphic function $E(s,f,h)$ extends meromorphically to all of $\mathbb{C}$, with only a finite number of poles in $Re(s) \geq 0$, all on the real axis. The connection with $L(s,\chi)$ may be explained by way of the operator $M(s,\chi)$ which intertwines $I(s,\chi)$ with $I(-s,\chi^{-1})$ and, at the same time, describes the "constant term" of $E(s,h)$. Given f in the space of $I(s,\chi)$ and $Re(s) \gg 0$, set

$$M(s,\chi)f(h) = \int_N f(w^{-1}nh)dn \ . \tag{I.2.1.3}$$

Here $N = \left\{\begin{bmatrix} 1 & x \\ 0 & 1 \end{bmatrix}\right\}$, $dn = dx$, and $w = \begin{bmatrix} 0 & 1 \\ -1 & 0 \end{bmatrix}$. The operator $M(s,\chi)$ clearly maps functions in the space of $I(s,\chi)$ to functions in the space of $I(-s,\chi^{-1})$. This operator may also be written as the product of local intertwining operators for $I(s,\chi_p)$ of the form

$$A(s,\chi_p)f_p(h) = \int_{N_p} f_p(w^{-1}nh)dn \tag{I.2.1.4}$$

where, for all p, $f_p \in I(s,\chi_p)$, and for almost all p, f_p is the unique $SL_2(\mathbb{Z}_p)$ fixed function f_p^0 equal to 1 at e. For any finite p, it is easy to compute that $w^{-1}n = n_1\begin{bmatrix} a & 0 \\ 0 & a^{-1} \end{bmatrix}k$, with $|a|_p^{-1} = max\{1,|x|_p\}$ if $n = \begin{bmatrix} 1 & x \\ 0 & 1 \end{bmatrix}$. This implies that the integral defining $A(s,\chi_p)$ converges for $Re(s) \gg 0$; moreover, if both f_p and χ_p are unramified (i.e. $f_p = f_p^0$), a simple computation gives

$$A(s,\chi_p)f_p(e) = \frac{1-\chi(p)p^{-(s+1)}}{1-\chi(p)p^{-s}} \ .$$

From this it follows that

$$M(s,\chi)f = \left(\prod_{p\in S} A(s,\chi_p)f_p\right)\left(\prod_{p\notin S} \tilde{f}_p^0\right)\frac{L_S(s,\chi)}{L_S(s+1,\chi)} \ , \tag{I.2.1.5}$$

where S denotes the set of primes p where χ_p is ramified (or $p = \infty$), $L_S(s,\chi) = \prod_{p \notin S}(1-\chi(p)p^{-s})^{-1}$, and $\tilde{f}_p^0$ denotes the unique K_p-invariant function in the space of $I(-s, \chi_p^{-1})$ such that $\tilde{f}_p^0(e) = 1$.

On the other hand, using the Bruhat decomposition $H = B \cup Bw^{-1}N$, we have

$$\text{(I.2.1.6)} \qquad \int_{N_{\mathbb{Q}} \backslash N_{\mathbb{A}}} E(s,f,nh)dn = f(h) + M(s,\chi)f(h) \; ,$$

i.e., $M(s,\chi)$ describes the "constant term" of (the automorphic form) $E(s,f,h)$ along N. Therefore, $M(s,\chi)$ also enjoys a meromorphic continuation to all of $\mathbb{C}$, and from (I.2.1.5) the promised meromorphic continuation of $L_S(s,\chi)$ is established.

Remark. The theory of Eisenstein series also provides a functional equation for $E(s,f,h)$, and this in turn gives some information on the functional equation for the *quotient* of L-functions $\frac{L_S(s,\chi)}{L_S(s+1,\chi)}$. More about this later.

2.2. L-groups.

The special example of $L(s,\chi)$ just described already exhibits several features of Langlands' general theory. If we view χ as an automorphic representation of $G = GL(1)$ (cf. [Ge2], Chapter 3), then we may view $L(s,\chi_p)$ as $L(s,\pi_p,r)$, where $\pi_p - \chi_p$ and r is the "standard" representation of ${}^LG = GL_1(\mathbb{C}) = \mathbb{C}^x$. Note that $G = GL(1)$ embeds as the "Levi-component" of the Borel (parabolic) subgroup of $H = SL(2)$ (the group on which our Eisenstein series are defined); moreover, the standard representation of $GL_1(\mathbb{C})$ happens to coincide with the natural adjoint action of ${}^LG$ on (the Lie algebra of) the unipotent radical of the parabolic subgroup ${}^LG^LN = \left\{ \begin{bmatrix} \alpha & y \\ 0 & 1 \end{bmatrix} \right\}$ of ${}^LH = PGL_2(\mathbb{C})$. As we shall see below, this too is a feature of the general theory.

We begin by recalling the general notion of L-group. Although we shall continue to focus on special examples as much as possible, there is no getting around the need to define certain notions (such as the L-group) in a completely general setting.

For G any connected reductive algebraic group over (the $\mathbb{A}$-field) F, with maximal torus T and associated "based root datum" $\psi_0(G,T) = (X^*(T), \Delta, X_*(T), \Delta^v)$, we let ${}^LG^0$ denote the (unique) complex reductive Lie group (with maximal torus ${}^LT)$) whose associated root datum is isomorphic to the"dual root datum" $\psi_0^v(G,T) = (X_*(T), \Delta^v, X^*(T), \Delta)$. For details on the correspondence $(G,T) \longrightarrow \psi_0(G,T) \longrightarrow {}^LG^0$, see Sections 1-3 of [Spr2], or Sections 9.1.4–9.1.6 and Chapter 12 of [Spr1]. It is important to recall at this point that ${}^LG^0$ depends only on G regarded as a group *over the algebraic closure* $\bar{F}$ *of* F; in particular, ${}^LG_1^0 \approx {}^L G_2^0$ whenever G_1 and G_2 are isomorphic over some finite Galois extension of F.

If $G = GL(n)$, then ${}^LG^0 = GL_n(\mathbb{C})$, since the based root datum $\psi_0(G)$ is self-dual. More generally, if G is of type A_n, B_n, C_n, D_n, E_6, E_7, E_8, F_4 or G_2 (in the Killing-Cartan classification) then it is known that the map $\psi_0(G) \longrightarrow \psi_0^v(G)$ permutes B_n and C_n and leaves all other types stable. (In particular, if $G = SO_{2n+1}$, then ${}^LG^0 = Sp_n(\mathbb{C})$.) For more details on this (and the definitions below) see [Bo1] and [Spr1]. The definition of ${}^LG^0$ of course also makes sense over any completion of the global field F. Henceforth, let us fix F to be such a local field, and consider G as a group over F, with L-group ${}^LG^0$. The significance of this L-group, roughly, is that equivalence classes of irreducible "admissible" representations of G (over a local field) should be indexed by homomorphisms of the local Galois group into ${}^LG^0$. This assertion is just one part of the vast "Langlands Program"; for $GL(n)$ it amounts to the conjecture that the irreducible representations of GL_n are naturally associated to n-dimensional representations of the local Galois group. (Actually, to be precise, we should be dealing with the "Weil-Deligne" group W_v' in place of the Galois group; cf. [Tate2] and [Bo1].)

For *unramified* representations, this correspondence is easier to describe. In this case, any "Langlands parameter" $\varphi: W_v' \longrightarrow {}^LG^0$ is determined by the image of any Frobenius automorphism, i.e., by $\varphi(Fr) = t$, where the element t in ${}^LG^0$ may be assumed to lie in ${}^LT^0$, and to be uniquely determined up to conjugacy. Thus the problem is to relate "semisimple" conjugacy classes in ${}^LG^0$ to unramified representations of G_v. It is this case which is crucial to the description of Langlands' automorphic L-functions, and which we describe below in some detail.

2.3. The unramified representations.

Fix F to be a non-archimedean p-adic field. We shall assume (as we have implicitly been doing all along) that our group G is split over F, i.e., its maximal torus T is defined over F, and isomorphic to $(F^x)^n$. We let $({}^LG^0, {}^LT^0)$ denote the corresponding "dual" group and maximal torus, and recall that $X^*({}^LT^0)$ is isomorphic to $X_*(T)$. The existence of a canonical correspondence

$$\text{(I.2.3.1)} \qquad \pi \longrightarrow t_\pi ,$$

pairing unramified representations of G with semisimple conjugacy classes in ${}^LG^0$ is based directly on the so-called *Satake isomorphism*. This isomorphism describes the spherical function algebra $\mathcal{H}(G,K)$ in terms of the affine algebra of the complex variety ${}^LT^0/W$. (Here K denotes a hyperspecial maximal compact subgroup of G, $\mathcal{H}(G,K)$ denotes the algebra of bi-K-invariant compactly supported "smooth" functions on G, and W denotes the Weyl group of ${}^LG^0$).

Using Satake's isomorphism, any unramified representation π may be shown to be a constituent of a "spherical principal series" representation $ind\,{}_B^G\chi$, where $B = TN$ is a Borel subgroup containing T, and χ is an unramified character of T (i.e., χ is trivial on the maximal compact subgroup ${}^0T$ of T); cf. §4 of [Car]. Establishing the correspondence (I.2.3.1) is thus reduced to the task of relating such unramified characters to t in ${}^LT^0$. To this end, consider the exact sequence

$$\text{(I.2.3.2)} \qquad 1 \longrightarrow {}^0T \longrightarrow T \stackrel{ord_T}{\longrightarrow} X_*(T) \longrightarrow 1$$

where the epimorphism $ord_T : T \longrightarrow X_*(T)$ is characterized by $\langle ord_T(t), \lambda\rangle = val_F(\lambda(t))$ for t in T, λ in $X^*(T)$, and $\langle\ ,\ \rangle$ the natural $\mathbb{Z}$-pairing on $X_*(T) \times X^*(T)$. Unramified characters naturally correspond to points in $Hom(X_*(T), \mathbb{C}^x)$, and these in turn correspond to points of the complex torus ${}^L T^0$ (since $X^*({}^L T^0) = X_*(T)$, and ${}^L T^0 = Spec\ \mathbb{C}[X_*(T)]$). In short, we have a well-defined isomorphism

$$t \longrightarrow \chi_t$$

between the points of ${}^L T^0$ and the group of unramified characters of T; moreover, two such characters χ_t and $\chi_{t'}$ will be in the same Weyl-group orbit iff t and t' are conjugate in ${}^L G^0$. The correspondence $t \longleftrightarrow \chi_t$ can be fixed by the relation

$$\chi_t(\varphi(\tilde{\omega}_F)) = \varphi(t) \tag{I.2.3.3}$$

for t in ${}^L T^0$, φ in $X_*(T) = X^*({}^L T^0)$, and $\tilde{\omega}_F$ any uniformizing variable in F. For $G = GL_n$ and $\pi = ind\ \mu_1\mu_2 \ldots \mu_n$, this relation gives

$$t = t_\pi = \begin{bmatrix} \mu_1(\tilde{\omega}) & & 0 \\ & \ddots & \\ 0 & & \mu_n(\tilde{\omega}) \end{bmatrix}. \tag{I.2.3.4}$$

<u>**The General Case.**</u> The above notions and results all hold more generally for *arbitrary reductive groups*, once ${}^L G^0$ is replaced by the larger L-group

$${}^L G = {}^L G^0 \rtimes \Gamma_F ,$$

with Γ some form of the Galois group of F acting on ${}^L G^0$. Here the action of Γ_F definitely depends on the "relative theory" for G, i.e., how G is defined over F (as opposed to $\bar{F}$). If G happens to be *split* over F, then this action of Γ_F is trivial, and we can replace ${}^L G$ by just ${}^L G^0$. More generally, if G_1 and G_2 are F-forms of each other, i.e., isomorphic to each other over $\bar{F}$, then ${}^L G_1 \approx {}^L G_2$ if and only if G_1 and G_2 are inner forms of each other (cf. §3 of [Spr2]).

The case of most interest for the theory of L-functions is the case when G is *quasi-split* over F, i.e.when G possesses a Borel subgroup defined *over* F. Indeed suppose just for the moment that F is a *global* $\mathbb{A}$-field and G is any connected reductive algebraic group which is defined over F, and splits over some finite Galois extension K of F. Then we consider the "finite form" of the L-group ${}^LG$ over F obtained by dividing Γ_F by the closed normal subgroup $Gal(\bar{F}/K)$ which acts trivially on ($\psi_0(G,T)$ and hence) ${}^LG^0$. Thus we replace Γ_F by the finite group $Gal(K/F)$, and consider the group

$$^LG = {}^LG^0 \rtimes Gal(K/F)$$

which is a complex reductive Lie group with connected component ${}^LG^0$. For almost every place v of F it is known (cf. §4 of [Spr2]) that G is quasi-split over F_v, and in fact splits over the unramified extension K_w (with w a place of K dividing v). In this case, the unramified representations of G_v are in 1-1 correspondence with ${}^LG^0$-semisimple conjugacy classes $t \times Fr_p$ in ${}^LG_v = {}^LG^0 \rtimes Gal(K_w/F_v)$, Fr_p a generator of $Gal(K_w/F_v)$. More precisely, Satake's isomorphism again associates to each such π the W-orbit of an unramified character χ of the maximal torus T (no longer necessarily split over F), and this χ now corresponds to the ${}^LG^0$-conjugacy class of a point $t \rtimes Fr_p$ in ${}^LG_v$.

Example (I.2.3.5): The quasi-split unitary group in 3 variables – (cf. [GePS1]).

Fix an unramified quadratic extension K of the local field F, and V a unitary 3 dimensional space over K. We pick a basis for V so the matrix corresponding to the hermitian form on V is $J = \begin{bmatrix} 0 & 0 & 1 \\ 0 & 1 & 0 \\ 1 & 0 & 0 \end{bmatrix}$, and let $G = U_{2,1}$ denote the corresponding (unitary) group of isometries, i.e., g in $GL_3(K)$ belongs to G if and only if $(gx, gy) = (x,y)$ for all x, g in K^3. (Here (,) denotes the *Hermitian* form defined on $K^3 \times K^3$ by $(x,y) = {}^t\bar{x}Jy$. Equivalently, g belongs to G if and only if ${}^t\bar{g}Jg = J$.)

To see that G is a (quasi-split) group defined *over* F, it is convenient to pick a basis $\{1, i\}$ for K over F such that $i^2 \in F$, and such that the non-trivial automorphism σ of $Gal(K/F)$ takes $x + iy$ in K to $x - iy$. If g in $GL_3(K)$ has the matrix form $g_1 + ig_2$ with respect to some basis k_1, k_2, k_3 of K^3, with g_1, g_2 in $GL_3(F)$, then it is easy to check that g

(regarded as a linear transformation of the six dimensional space K^3 *over* F) has the matrix form

$$\begin{pmatrix} g_2 & i^2 g_2 \\ g_2 & g_1 \end{pmatrix}$$

with respect to the basis $\{k_1, k_2, k_3, ik_1, ik_2, ik_3\}$. Moreover, the original invertible matrix belongs to G (i.e. ${}^t\bar{g}Jg = J$) if and only if the matrix $\begin{pmatrix} g_1 & i^2 g_2 \\ g_2 & g_2 \end{pmatrix}$ in $GL_6(F)$ satisfies

$$(*) \quad \begin{pmatrix} g_1^t & -i^2 g_2^t \\ -g_2^t & {}^t g_1 \end{pmatrix} \begin{pmatrix} J & 0 \\ 0 & J \end{pmatrix} \begin{pmatrix} g_1 & i^2 g_2 \\ g_2 & g_1 \end{pmatrix} = \begin{pmatrix} J & 0 \\ 0 & J \end{pmatrix}.$$

Note that $(*)$ gives polynomial conditions on the coefficients of g_1, g_2 *defined over* F, i.e., G is indeed defined over F. On the other hand, as a group *over* K, $G \approx GL_3$ via the isomorphism

$$\begin{pmatrix} g_1 & i^2 g_2 \\ g_2 & g_1 \end{pmatrix} \overset{\varphi}{\longleftrightarrow} g_1 + ig_2 \ .$$

Thus ${}^LG^0$ is just $GL_3(\mathbb{C})$.

We claim next that ${}^LG = GL_3(\mathbb{C}) \rtimes Gal(K/F)$, where

$$\sigma(g) = \begin{bmatrix} 0 & 0 & 1 \\ 0 & 1 & 0 \\ 1 & 0 & 0 \end{bmatrix} {}^t g^{-1} \begin{bmatrix} 0 & 0 & 1 \\ 0 & 1 & 0 \\ 1 & 0 & 0 \end{bmatrix}$$

for each g in $GL_3(\mathbb{C})$; moreover, the conjugacy class in ${}^LG$ corresponding to the unramified representation π_ν induced from the character

$$\begin{bmatrix} a & * & * \\ 0 & b & * \\ 0 & 0 & \bar{a}^{-1} \end{bmatrix} \mapsto \nu(a)$$

of the Borel subgroup of G is simply

$$t_\nu = \begin{bmatrix} \nu(\tilde{\omega}) & 0 & 0 \\ 0 & 1 & 0 \\ 0 & 0 & 1 \end{bmatrix} \rtimes \sigma \ .$$

(Note that σ is just the Frobenius automorphism at F.)

We first describe the L-group by making explicit the action of $Gal(K/F)$ on the simple roots and, more generally, on $\psi_0(G)$. For a diagonal element $(a, b, \bar{a}^{-1})$ in the maximal torus T of B, write $a = a_1 + \imath a_2$, and $b = b_1 + ib_2$, where i (as before) is such that the non-trivial element σ of $Gal(K/F)$ takes i to $-i$, and $(b_1 + ib_2)(b_1 - ib_2) = 1$. The simple roots of G with respect to T are $\alpha_1(a, b, \bar{a}^{-1}) = (a_1 + ia_2)(b_1 - ib_2)$, and $\alpha_2(a, b, \bar{a}^{-1}) = (b_1 + ib_2)(a_1 - ia_2)$. Then we compute

$$\begin{aligned}(\sigma\alpha_1)(a, b, \bar{a}^{-1}) &= (a_1 - ia_2)(b_1 + ib_2) \\ &= \alpha_2(a, b, \bar{a}^{-1}) ,\end{aligned}$$

and similarly, $\sigma\alpha_2 = \alpha_1$. On the other hand, the automorphism $\sigma(g) = w^t g^{-1} w$ described above clearly induces this same automorphism of $\psi_0({}^L G^0)$. Thus our claim about ${}^L G$ follows from the well-known (split) exactness of the sequence

$$1 \longrightarrow Int(G) \longrightarrow Aut(G) \longrightarrow Aut\ \psi_0(G, T) \longrightarrow 1 ;$$

cf. §2.3 of [Bo1].

For the second part of the claim, we return to the sequence (I.2.3.2). When T is no longer the *split* maximal torus in G, the ord map is no longer always onto. Thus we let $\Lambda = ord(T)$, and we consider the exact sequence

$$1 \longrightarrow {}^0 T \longrightarrow T \xrightarrow{ord} \Lambda \longrightarrow 1 .$$

We note that $\Lambda = X_*(T)^{\Gamma_F} = X_*(T_d)$, where $X_*(T)^{\Gamma_F}$ denotes the fixed point set of the Galois group of F in $X_*(T)$, and T_d denotes the maximal *split* torus in T. Here in our example of $U_{2,1}$,

$$T_d = \left\{ \begin{pmatrix} x & & \\ & 1 & \\ & & x^{-1} \end{pmatrix} x \in F \right\} , \quad \text{and}$$

$$\Lambda \approx \mathbb{Z} \approx \left\{ x \longrightarrow \begin{pmatrix} x^n & & \\ & 1 & \\ & & x^{-n} \end{pmatrix} \right\} .$$

For $U_{2,1}$, this correspondence clearly attaches to the character $\nu\begin{pmatrix}\alpha & & \\ & \beta & \\ & & \bar{\alpha}^{-1}\end{pmatrix} = \nu(\alpha)$ the point $\nu(\tilde{\omega})$ in ${}^L T_d^0 \approx \mathbb{C}^x$. It remains to note that the natural inclusion of $\Lambda = X_*(T_d) = X^*({}^L T_d^0)$ into $X_*(T) = X^*({}^L T^0)$ induces the surjective morphism $i^* : {}^L T^0 \longrightarrow {}^L T_d^0$ given by

$$i^*\begin{pmatrix}a & & \\ & b & \\ & & c\end{pmatrix} = ac^{-1} .$$

Moreover, according to Lemma 6.4 of [Bo1], the map $t \rtimes \sigma \longrightarrow i^*(t)$ induces a bijection

$${}^L T^0 \rtimes \sigma \approx {}^L T_d^0 / {}_k W \ ,$$

where ${}_k W$ is the subgroup of the Weyl group W of (G,T) leaving T_d stable, and w in ${}_k W$ acts on ${}^L T^0 \rtimes \sigma$ via the formula $w(t \rtimes \sigma) = Int\ n(t \rtimes \sigma) = (n^{-1} \cdot t \cdot n^{\sigma}) \rtimes \sigma$, with n the inverse image of w the normalizer of ${}^L T^0$ in ${}^L G^0$. Thus our unramified representation $Ind\ \nu$ indeed determines the ${}^L G^0$ conjugacy class $\begin{pmatrix}\nu(\tilde{\omega}) & & \\ & 1 & \\ & & 1\end{pmatrix} \rtimes \sigma$ in ${}^L G$.

2.4. The general set-up: Preliminary definitions.

We return at last to Langlands' general theory, and make the following assumption on our reductive group G.

Assumption: G embeds as the Levi component of an F-parabolic subgroup P in a larger reductive group H, i.e., $H \supset P$ and P has Levi decomposition GU^P; moreover, we suppose that P is *maximal* among the F-parabolics of H.

Remarks. (i) These assumptions clearly generalize the set-up discussed in §I.(2.1) for obtaining the L-functions $L(s,\chi)$ from the theory of Eisenstein series on $H = SL(2)$. Thus it should not seem surprising now that certain automorphic L-functions $L(s,\pi,r)$ attached to G as above may also be studied by way of Eisenstein series on H induced from cusp forms on GU^P.

(ii) In general, we obtain groups G satisfying Hypothesis I by taking the Dynkin diagram of a reductive group H, deleting *one* of its simple roots, and considering the maximal parabolic subgroup P with Levi component corresponding to the resulting Dynkin diagram of one lower rank. For example, take H to be a simple group of type G_2, with Dynkin diagram $0\equiv\!\!\equiv 0$, and P to be the parabolic subgroup corresponding to the diagram 0, with Levi component of type A_1. Note that a group of type G_2 itself can *never* satisfy Assumption I, since there is a no larger Dynkin diagram containing $0\equiv\!\!\equiv 0$; a similar conclusion holds for the exceptional groups of type E_8 and F_4, and it is also clear that these are the *only* groups G which do not so imbed.

To explain the theory of [La1] in detail, and in particular to describe which $L(s,\pi,r)$ (i.e. which representations r) arise in this context, we need yet a few more definitions. For convenience, we reformulate the results of [La1] from the slightly more general point of view of [Sha 1,3,8]. Hence G is not necessarily split, or semisimple, or of adjoint type.

Our main task is to generalize and "normalize" the quasi-character $|a|^s_{\mathbb{A}}$ which appears in the description of the P-induced representation (I.2.1.2).

Suppose that H contains a Borel subgroup B also defined over F, and write $B = TU$, with T a maximal torus (also defined over F). Assume that our maximal parabolic subgroup $P = GU^P$ contains B, and fix a special maximal compact subgroup $K^H_v \subset H_v$ so that $H = BK^H = PK^H$ with $K^H = \Pi K^H_v$. Let A denote the maximal split torus in the center of G, with real Lie algebra

$$\mathcal{U} = Hom(X(G)_F, \mathbb{R}) = Hom(X(A)_F, \mathbb{R}) \ .$$

(Here $X(G)$ denotes the group of F-rational characters of G.) Now consider the homomorphism $H_P : G_{\mathbb{A}} \longrightarrow \mathcal{U}$ defined by

$$exp(H_P(g)(\chi)) = \prod_v |\chi(g_v)|_v$$

where $\chi \in X(G)_F$, and $g = (g_v) \in G_{\mathbb{A}}$. We extend H_P to $H_{\mathbb{A}}$ by making it trivial on $U^P_{\mathbb{A}}$ and $K^H_{\mathbb{A}}$.

Example (I.2.4.1) Let $H = SL(2) \supset B = P = TU$ with $T(= G) = \left\{\begin{bmatrix} a & 0 \\ 0 & a^{-1}\end{bmatrix}\right\} = A$. Then $X(G)$ is the group of maps $\chi_\nu\left(\begin{bmatrix} a & 0 \\ 0 & a^{-1}\end{bmatrix}\right) = a^\nu$, where ν ranges over $\mathbb{Z}$, the space $\mathcal{U}$ is isomorphic to $\mathbb{R}$, and

$$H_P\left(\begin{bmatrix} a & 0 \\ 0 & a^{-1}\end{bmatrix}\right)(\chi_\nu) = \nu\ log|a|_{\mathbb{A}}\ ,$$

i.e.,

$$H_P\left(\begin{bmatrix} a & 0 \\ 0 & a^{-1}\end{bmatrix}\right) = log\ |a|_{\mathbb{A}}\ .$$

Example (I.2.4.2) Consider $H = SL(3)$ and the parabolic subgroup

$$P = \left\{\begin{bmatrix} * & * & * \\ * & * & * \\ 0 & 0 & * \end{bmatrix}\right\} = GU^P$$

with

$$G = \left\{\begin{bmatrix} * & * & 0 \\ * & * & 0 \\ 0 & 0 & * \end{bmatrix}\right\} \approx GL(2)\ .$$

Then $X(G)_{\mathbb{Q}}$ is the group of maps

$$\chi_\nu\left(\left[\begin{array}{c|c} m & \begin{matrix}0\\0\end{matrix} \\ \hline 0\ 0 & (det\ m)^{-1}\end{array}\right]\right) = (det\ m)^\nu$$

where ν again ranges over $\mathbb{Z}$,

$$\mathbb{A} = \left\{\begin{bmatrix} a & 0 & 0 \\ 0 & a & 0 \\ 0 & 0 & a^{-2} \end{bmatrix}\right\},$$

the space $\mathcal{U}$ is again isomorphic to $\mathbb{R}$, and

$$H_P\left(\left[\begin{array}{c|c} m & \begin{matrix}0\\0\end{matrix} \\ \hline 0\ 0 & (det\ m)^{-1}\end{array}\right]\right) = log|det\ m|_{\mathbb{A}}\ .$$

Returning to the general set-up, let A_0 denote the maximal F-split torus in T, $\Phi = \Phi(H, A_0)$ the set of F-roots of A_0, Φ^+ the set of positive roots, i.e., those generating U (the unipotent radical of B), and $\Delta \subset \Phi^+$ the simple roots. Recall that A denotes the maximal split torus in the center of G. We identify the roots of A in U^P with a subset of Φ^+; in particular, the unique *reduced* root of A in U^P can be identified with an element in Δ which we henceforth call α. Let ρ_P denote half the sum of the F-roots generating U^P, and let $\langle\ ,\ \rangle$ denote the usual (Killing-Cartan) pairing between roots. Then $\tilde{\alpha} = \langle \rho_P, \alpha\rangle^{-1}\rho_P$ belongs to the dual space $\mathcal{U}^* = X(G) \otimes_{\mathbb{Z}} \mathbb{R} = X(A) \otimes_{\mathbb{Z}} \mathbb{R}$, and we shall identify $\mathbb{C}$ with a subspace of the complex dual $\mathcal{U}^*_{\mathbb{C}} = \mathcal{U}^* \otimes_{\mathbb{R}} \mathbb{C}$ via the identification $s \longrightarrow s\tilde{\alpha} \in \mathcal{U}^*_{\mathbb{C}}$.

Example. Consider $H = SL(2) \supset P = B = \left\{\begin{pmatrix} a & x \\ 0 & a^{-1}\end{pmatrix}\right\}$, with $T = A = A_0$, and Φ equal to the usual roots $\pm\alpha\begin{pmatrix} a & 0 \\ 0 & a^{-1}\end{pmatrix} = a^{\pm 2}$. Note $\rho_P = \frac{1}{2}\alpha$, and yet $\tilde{\alpha} = \frac{1}{2}\alpha = \rho_P$ since $\langle \rho_P, \alpha\rangle = 1$. Thus $|a|^s_{\mathbb{A}} = exp(s\ log|a|_{\mathbb{A}}) = exp(H_P\begin{pmatrix} a & 0 \\ 0 & a^{-1}\end{pmatrix})(s\tilde{\alpha}) \equiv exp(H_P\begin{pmatrix} a & 0 \\ 0 & a^{-1}\end{pmatrix}(s))$.

We now define a character of $G_{\mathbb{A}}$ (the natural generalization of $|a|^s_{\mathbb{A}}$ on $G_{\mathbb{A}}$) by the formula

$$g \longrightarrow exp(H_P(g)(s)) = \prod_v e^{H_{P_v}(g_v)(s)}$$

where $e = q_v$ (the residual characteristic of F_v) if $v < \infty$, s is identified with $s\tilde{\alpha}$ in $\mathcal{U}^*_{\mathbb{C}}$, and we have extended the pairing on $X(G) \times \mathcal{U}$ to $\mathcal{U}^*_{\mathbb{C}} \times \mathcal{U}_{\mathbb{C}}$ in the natural way.

2.5. The general set-up: Eisenstein series, constant terms, and Langlands' "Euler Products".

Fix $\pi = \otimes\pi_v$ an automorphic cuspidal representation of $G_{\mathbb{A}}$. By this we mean a collection of irreducible unitary representations π_v of the local groups G_v satisfying local and global conditions completely analogous to those given in §I.1.1 (see also Sections 13, 14 of [Bo1]). For s in $\mathbb{C}$, consider the induced representation

$$(\text{I.2.5.1}) \qquad I(s,\pi) = Ind_{P_{\mathbb{A}}}^{H_{\mathbb{A}}} \pi \otimes exp(H_P(g)(s)) \otimes 1 \ .$$

For $G = GL(1) \subset B = \left\{ \begin{pmatrix} a & x \\ 0 & a^{-1} \end{pmatrix} \right\} \subset H = SL_2$, and $\pi = \chi$, this reduces to the representation $Ind_{B_{\mathbb{A}}}^{H_{\mathbb{A}}} \chi(a)| \, |_{\mathbb{A}}^s = I(s,\chi)$ of §I.(2.1). In general, recall that $P_{\mathbb{A}} = G_{\mathbb{A}} U_{\mathbb{A}}^P$, and the 1 in the formula above imples that $\pi \otimes exp(H_P(g)(s))$ is extended trivially across $U_{\mathbb{A}}^P$. Note that $I(s,\pi) = \otimes I(s,\pi_v)$ with

$$I(s,\pi_v) = Ind_{P_v}^{H_v} \pi_v \otimes q_v^{H_P(\cdot)(s)} \otimes 1$$

(and q_v replaced by exp if $v = \infty$).

An *Eisenstein series on* $H_{\mathbb{A}}$ is defined by

$$(\text{I.2.5.2}) \qquad E(s,f,P.h) = \sum_{\gamma \in P_F \backslash H_F} f_s(\gamma h)$$

where f_s is a "nice" function in the space of $I(s,\pi)$. In particular, we require that f_s be a continuous right $K_{\mathbb{A}}$-finite function on $H_{\mathbb{A}}$ which is left $U_{\mathbb{A}}^P$-invariant and such that for each h in $H_{\mathbb{A}}$

$$g \longrightarrow f_s(gh)$$

defines a cusp form on $G_{\mathbb{A}}$ belonging to the space of the irreducible cuspidal representation $\pi \otimes exp(H_P(g)(s)) \otimes \delta_P^{1/2}(g)$ of $G_{\mathbb{A}}$. Here δ_P denote the modulus function of P, i.e. the ratio of right to left Haar measures on P.

Remark. The relation

$$\delta_P(p) = exp(H_P(p)(2\rho_P)) \ ,$$

suggests that we define the quasi-character $|p|_{\mathbb{A}}^s$ directly in terms of this modulus function, say by

$$|p|_{\mathbb{A}}^s = (\delta_P(p))^{s/2} \ .$$

This definition has the advantage of avoiding the function $H_P(\cdot)$, and is essentially the approach taken in §5 of [La1]. However, our definition of $|\ |_{\mathbb{A}}^s$ on $G_{\mathbb{A}}$ has already built into it the proper "normalization" of s which (as we shall soon see) simplifies the natural generalization of (I.2.1.5).

As in the example of $H = SL(2)$, there will be an intertwining operator

$$M(s,\pi) = \bigotimes_v A(s,\pi_v)$$

which appears in a constant term of $E(s,f,P,h)$. We shall not now explain which induced representations of H this operator intertwines, except to say that the operator itself corresponds to the Weyl group element w of H taking α to a negative root and the remaining simple roots into Δ.

It is a consequence of the general theory of Eisenstein series ([La3]) that $M(s,\pi)$ extends to a meromorphic function of s in $\mathbb{C}$, with only a finite number of poles in the plane $Re(s) \geq 0$, all simple and on the real axis.

Caution. We know little about the poles of $M(s,\pi)$ (or $E(s,f,P,h)$) in the *left* half-plane$Re(s) \leq 0$. Cases where we *can* prove the finiteness of poles in the whole s-plane will be discussed in the Appendix A.2 of Section II.1.2 and in Section III.1.5, and will of course play a crucial role in the theory of L-functions.

Now we describe the generalization of relation (I.2.1.5) which expresses $M(s,\pi)$ in terms of certain automorphic L-functions. Let ${}^LP$ denote the parabolic subgroup of ${}^LH$ corresponding to P in H, and let ${}^L\eta$ denote the Lie algebra of the unipotent radical of ${}^LP$. The group ${}^LG$ acts on ${}^L\eta$ via the adjoint action, and the corresponding representation $r : {}^LG \longrightarrow GL({}^L\eta)$ decomposes into the sum of certain "eigenvalue" representations r_i, $i = 1, \ldots, m$. For $P = B \subset H = SL_2$, this representation r is itself already irreducible; in particular, $m = 1$ and $r = r_1$.

In general, we let r_i denote the restriction of r to the space of all dual root vectors $X_{\beta^\vee}$ in ${}^L\eta$ such that $\langle\tilde{\alpha},\beta\rangle = i$. Recall that $\tilde{\alpha} = \langle\rho_P,\alpha\rangle^{-1}\rho_P \subset \mathcal{U}^* = X(G) \otimes_Z \mathbb{R}$, so $\langle\tilde{\alpha},\beta\rangle$ denotes the usual pairing on roots, extended linearly. For example, in case $H = SL(2) \supset P = AN$, the one-dimensional space ${}^L\eta$ is spanned by $X_{\beta^v} = \begin{bmatrix} 0 & 1 \\ 0 & 0 \end{bmatrix}$, with β^v the co-root corresponding to the root $\beta = \alpha$; in this case, $\tilde{\alpha} = \rho_P = \frac{1}{2}\alpha = \frac{1}{2}\beta$, and $\langle \tilde{\alpha}, \beta \rangle = 1$. In general, the subrepresentations r_i so defined are irreducible (at least when the Dynkin diagram of H is irreducible); cf. Section 4 of [Sha8].

The thrust of the calculations in [La1] and [Lai] is that

$$
\text{(I.2.5.3)} \qquad \begin{aligned} M(s,\pi)f = &\left(\bigotimes_{v\in S} A(s,\pi_v)f_v\right) \otimes \bigotimes_{v\notin S} \tilde{f}_v^0 \\ &\prod_{i=1}^{m} \frac{L_S(is,\pi,\tilde{r}_i)}{L_S(1+is,\pi,\tilde{r}_i)}, \end{aligned}
$$

for $Re(s)$ sufficiently large (depending on the rank of H). Here S denotes a finite set of places of F (including all the archimedean ones) such that for every $v \notin S$, π_v and H_v are both "unramified", $f = \otimes_v f_v$ is such that for each $v \notin S$, f_v is the unique $K_v^{H_v}$-fixed function with $f_v^0(e_v) = 1$, $\tilde{f}_v^0$ is the analogous $K_v^{H_v}$-fixed vector in the image of the intertwining operator $A(s,\pi_v)$, $\tilde{r}_i$ denotes the contragredient of r_i, and

$$
L_S(s,\pi,\tilde{r}_i) = \prod_{v\notin S} L(s,\pi_v,\tilde{r}_{i,v})
$$

with $L(s,\pi_v,\tilde{r}_{i,v})$ the local Langlands factor

$$
\{det[I - \tilde{r}_{i,v}(t_v)q_v^{-s}]\}^{-1} .
$$

Remarks (I.2.5.4) (i) By a representation of ${}^LG$ we mean a continuous homomorphism ${}^LG \longrightarrow GL_m(\mathbb{C})$ whose restriction to ${}^LG^0$ is a morphism of complex Lie groups. Because of the natural homomorphism of $Gal(K_w/F_v)$ into $Gal(K/F)$, there exist canonical homomorphisms $\eta_v : {}^LG_v \longrightarrow {}^LG$ so that each representation r_i above defines one of ${}^LG_v$ by composition with η_v.

(ii) The semisimple conjugacy class t_v appearing in (2.54) is *no longer* fixed by the rule (I.2.3.3), but rather (as in [La1]) by the relation $\chi_t(\varphi^{-1}(\tilde{\omega})) = \varphi(t)$; this explains why the *contragredient* of r_i appears in (I.2.5.3), and why (2.5.3) is indeed consistent with the formula (2.1.5) which it so vastly generalizes.

Note that (I.2.5.3) establishes the meromorphic continuation of the automorphic L-functions $L(s, \pi, \tilde{r}_1)$ if, for example, $m = 1$, or the remaining L-functions are known. In particular, we obtain a meromorphic continuation for the (partial) Jacquet-Langlands L-function $L(s, \pi)$ by applying (I.2.5.3) in the context of Example (2.4.2), where $H = SL_3$, $G = GL(2)$, and π is any automorphic cuspidal representation of $G_{\mathbb{A}}$. Because $Ad(^LG)$ acts irreducibly on the Lie algebra of

$$U^P = \left\{ \left(\left[\begin{array}{c|c} I_2 & \begin{matrix} x \\ y \end{matrix} \\ \hline 0\ 0 & 1 \end{array} \right] \right) \right\} \subset \ ^LH = PGL_3(\mathbb{C})$$

via the standard representation $\rho_2 : GL_2(\mathbb{C}) \longrightarrow GL_2(\mathbb{C})$, $L_S(s, \pi, r_1)$ indeed equals $L_S(s, \pi)$; unfortunately, no information on the holomorphy (or exact functional equation) of $L(s, \pi)$ is thus obtained.

In general, the question arises: which groups G can be embedded as the Levi component of a maximal F-parabolic subgroup of a reductive group H, and what are the corresponding possible representations r_i? If (G, π, r) is an L-function datum such that G and r can be so embedded, we call this datum *admissible.* As already remarked at the beginning of Section 2.4, we know that all but three split adjoint groups (namely G_2, E_8 and F_4) are "admissible", whereas a long list of possible corresponding r_i's is described in Chapter 6 of [La1]. We shall soon discuss several important examples from this list, as well as from the completion of this list (for quasi-split as well as split groups) to be found in [Sha8]. In quoting from these lists, it is important to keep in mind the following:

(1) Our subscript m (labelling the last of the representations $r_1, \ldots, r_m$) appears as the integer r in [La1], and our representations r_i

appear as his representations π_i.

(2) Langlands' eigenvalues for π_i (cf. p.45 of [La1]) appear as integers a_i, whereas our identification of s in $\mathbb{C}$ with $s\tilde{\alpha}$ in $\mathcal{U}^*_{\mathbb{C}}$ was fixed *precisely* so that $a_i = i$.

(3) To be consistent with [Sha8], we order our root subscripts according to the roots of H (rather than G as in [La1]).

For example, consider the diagram

(ii)

	$0\!-\!\!-\!0 \ldots 0\!-\!\!-\!0$	
	$\alpha_1 \quad \alpha_2 \quad\quad \alpha_{n-1}$	
$m(=r) = 1$	$(a_1 = n+1)$	$\lambda_1 = \delta_1$

which is found on page 47 of [La1]. (We have replaced n by $n-1$, as explained in (3) above. When $n = 2$, this example corresponds to the case $GL_2 \subset SL_3$ discussed above. In general, the full Dynkin diagram corresponds to the group $H = SL_{n+1}$, and the diagram determined by the *labelled* simple roots corresponds to the Levi component (GL_n) of the parabolic subgroup P of H. The representation $r_1 = r$ is the standard representation $\rho_n : GL_n(\mathbb{C}) \longrightarrow GL_n(\mathbb{C})$ with highest weight λ_1, equal to the fundamental weight δ_1.

A slightly different example is (iii) on page 47 of [La1]:

$$\begin{array}{ccccc} 0\!-\!\!-\!0 & \ldots & 0\!-\!\!-\!0\!-\!\!-\!0 & \ldots & 0\!-\!\!-\!0 \\ \alpha_1 & & \alpha_{m-1} \quad \alpha_{m+1} & & \alpha_{n-1} \quad \alpha_n \\ r = 1 & & & & \lambda_1 = \delta_1 + \delta_n \ . \end{array}$$

Here $H = GL_{n+1}$, G can be taken to be the group $GL_m \times GL_{n-m+1}$, and $r : GL_m(\mathbb{C}) \times GL_{n-m+1}(\mathbb{C}) \longrightarrow GL_{m(n-m+1)}(\mathbb{C})$ is the tensor product of the standard representation of $GL_m(\mathbb{C})$ with the contragredient of the standard representation of $GL_{n-m+1}(\mathbb{C})$. (The highest weight of this representation is $\delta_1 + \delta_n$.)

As we shall explain in Chapter III, [Sha8] completes Langlands' list in the split case, and treats the quasi-split (reductive) case as well. In these latter lists, the abundance of examples with *non*-simple G provides the rationale for ordering the roots of the Dynkin diagram according to the roots of H rather than G.

CHAPTER II.

DEVELOPMENTS AND REFINEMENTS (1970-1982)

§1. ZETA-INTEGRALS FOR GL(n) AND BEYOND

In §1 of Chapter I we explained how local and global zeta-integrals are used to prove the analytic continuation and functional equations of the automorphic L-functions attached to $GL(2)$ (Steps 1 through 5 of Section I.1.7). In the decade immediately following the work of Jacquet and Langlands, similar strategies were developed for $GL(n)$, groups closely related to $GL(n)$, and representations r of ${}^LG$ closely related to the standard representation. Since the purpose of Steps (1)-(5) should by now be clear to the reader, our task here will be to emphasize the main new ideas introduced in each of the special cases discussed.

1.1. The method of Tate-Godement-Jacquet.

In §14 of [JL] a second kind of zeta-integral was introduced which does not rely on the theory (or existence) of Whittaker models, and therefore applies just as well to the case of the multiplicative group of a division quaternion algebra (where there are no unipotent elements like $\begin{bmatrix} 1 & x \\ 0 & 1 \end{bmatrix}$). This is the idea developed more generally in [Go Ja] for the multiplicative group G of the matrix algebra $M_n(D)$, where D is a (central) division algebra over F. For simplicity of exposition, we shall assume $G = GL_n(F)$.

First suppose F is a local non-archimedean field, and π is an irreducible "admissible" representation of G on the complex vector space V. There is then a (unique) irreducible admissible representation "contragredient" (or dual) to π acting on a space V, and a non-degenerate bilinear form $\langle \cdot, \cdot \rangle$ on $V \times \tilde{V}$ which is G-invariant. (See Section 1 of [Go Ja] for a careful discussion of these notions.) Every function f of the form

$$f(g) = \langle \pi(g)v, \tilde{v} \rangle \ , \quad v \in V \ , \quad \tilde{v} \in \tilde{V} \ , \qquad \text{(II.1.1.1)}$$

is called a *matrix coefficient* of π. A *zeta-integral* attached to (π, V) is defined through the formula

$$\text{(II.1.1.2)} \qquad \mathcal{Z}(s, f, \Phi) = \int_{M_n(F)} \Phi(x) f(x) |det(x)|^s d^x x$$

where f is any matrix coefficient of π, and Φ is any Schwartz-Bruhat function on $M_n(F)$. (A Schwartz-Bruhat function on $M_n(F)$ is any complex-valued locally constant compactly supported function, and the totality of such functions is denoted $\mathcal{S}(M_n(F))$.)

In case $n = 1$, any irreducible admissible representation of $GL_1(F) = F^x$ is a quasicharacter of the form $\chi(x)$, and so $\mathcal{Z}(s, f, \Phi)$ reduces to the familiar zeta-integral

$$\text{(II.1.1.3)} \qquad \mathcal{Z}(s, \chi, \Phi) = \int_F \Phi(x)\chi(x)|x|^s d^x x$$

analyzed in [Tate1]. The method of [Go Ja] (and Section 14 of [JL]) is thus a natural generalization of the method used by Tate to establish the analytic continuation and functional equation of Hecke's L-series $L(s, \chi)$. The idea of this generalization was apparently first (systematically) developed by Tamagawa and Godement in the 1960's, but the method is still often referred to as "Tate's method" (even in the general context of $M_n(D)$).

In general, suppose (π, V) is an *unramified* representation of $GL_n(F)$, $\Phi(x)$ is equal to the characteristic function of $M_n(O_F)$, and $f(x) = \langle \pi(g)v_0, \tilde{v}_0 \rangle$ with v_0 (resp. $\tilde{v}_0$) the $GL_n(O_F)$ fixed vector in V (resp. $\tilde{V}$) such that $f(e) = 1$. Then a pleasant computation shows that

$$\mathcal{Z}(s, f, \Phi) = det[I - \rho_n(t_\pi)q_v^{-s}]^{-1}$$

where $\rho_n : {}^L G \longrightarrow GL_n(\mathbb{C})$ is the standard representation, and t_π in $GL_n(\mathbb{C})$ is the conjugacy class in ${}^L G = GL_n(\mathbb{C})$ corresponding to π (cf. I.2.3.4). The global analogue of (II.1.1.2) may be defined for any automorphic cuspidal representation $\pi = \otimes \pi_v$ of $GL_n(\mathbb{A}_F)$ by way of the integral

$$\text{(II.1.1.4)} \qquad \mathcal{Z}(s, \varphi, \Phi) = \int_{G_{\mathbb{A}}} \Phi(x)\varphi(x)|det(x)|_{\mathbb{A}}^s d^x x \,.$$

Here Φ is $\mathcal{S}(M_n(\mathbb{A}))$ is of the from $\Pi\Phi_v(x)$, where $\Phi_v \in \mathcal{S}(M_n(F_v))$ is almost everywhere equal to the characteristic function of $M_n(O_{F_v})$, and φ is any cusp form in the subspace of $L^2(G_F\backslash G_{\mathbb{A}},\omega)$ realizing π (in particular, φ transforms under the center of $GL(n)$ according to the "grossencharacter" ω). With more difficulty than in the case $n = 1$ (or the case $n = 2$ in [JL]) it can be shown that $Z(s,\varphi,\Phi)$ – initially defined only in some right half-plane by a convergent integral – extends analytically to the whole complex plane as an entire function, and satisfies the (global) functional equation

$$Z(s,\varphi,\Phi) = Z(n-s,\check{\varphi},\hat{\Phi}) \; . \tag{I.1.1.5}$$

Here $\hat{\Phi}$ denotes the Fourier transform $\int_{\mathbb{A}} \Phi(y)\psi(yx)dy$, and $\check{\varphi}(g) = \varphi(g^{-1})$; for details, see Sections 11-13 of [Go Ja].

More precisely, as we wish to obtain an Euler product expansion in terms of the local zeta-integrals $Z(s,f_v,\Phi_v)$, we deal not really with (II.1.1.4), but rather with a global zeta-integral of the form

$$Z(s,f,\Phi) = \int_{G_{\mathbb{A}}} \Phi(x)f(x)|det\ x|^s_{\mathbb{A}}dx \tag{II.1.1.6}$$

where

$$f(g) = \int_{Z_{\mathbb{A}}G_F\backslash G_{\mathbb{A}}} \varphi(hg)\tilde{\varphi}(h)dh \tag{II.1.1.7}$$

is the matrix coefficient of $\pi = \otimes\pi_v$ in $L^2_0(G_F\backslash G_{\mathbb{A}},\omega)$ corresponding to φ in $H_\pi \subset L^2_0(\omega)$ and $\tilde{\varphi}$ in $H_{\tilde{\pi}} \subset L^2_0(\omega^{-1})$. Steps 1 and 2 of §I.(1.7) then follow as discussed above, and it remains to verify the local properties required in Steps (3)-(5).

Outside of the "unramified case" (which almost always provides a pleasant respite from the general theory), the functional equation and detailed properties of the local zeta-integrals $Z(s,f_v,\Phi_v)$ are not so easy to establish. First Jacquet's theory of "Jacquet modules" is used to reduce matters to the case when π_v is not contained in any representation induced from a proper parabolic subgroup, and then the resulting "supercuspidal" π_v are treated separately.

It should be stressed again that the local factors $L(s, \pi_v)$ are introduced as g.c.d.'s of the zeta-integrals $Z(s, f_v, \Phi_v)$ for any irreducible admissible representation π_v; therefore the Euler product

$$L(s, \pi) = \prod_v L(s, \pi_v)$$

makes sense for *any* irreducible unitary representation $\pi = \otimes \pi_v$ of $G_{\mathbf{A}}$. However, the proof that $L(s, \pi)$ is "nice" (has an analytic continuation and functional equation) starts from the fact that π is *automorphic cuspidal.* Indeed, only in this case are the appropriate zeta-integrals even defined.

1.2. Jacquet's theory for $GL(2) \times GL(2)$ and the method of Rankin-Selberg.

[Jacquet] is a 1972 sequel to [JL], which analyzes the L-functions $L(s, \pi_1 \times \pi_2)$ attached to *pairs* of automorphic cuspidal representations of $GL(2)$; in this work, the classical papers of [Rankin1] and [Selberg] play the same role as Hecke's work on the Dirichlet series $L(s, f)$ plays in [JL]. Thus the method developed in [Jacquet] is often referred to as the "Rankin-Selberg method".

Classically, the goal is to study analytic properties of the Dirichlet series

$$\sum_{n=1}^{\infty} \frac{a_n \bar{b}_n}{n^s} ,$$

when $\sum_{n=1}^{\infty} a_n e^{2\pi i n z}$ and $\sum_{n=1}^{\infty} b_n e^{2\pi i n z}$ define automorphic forms of weight k for $SL_2(\mathbb{Z})$. The "basic identity" exploited by Rankin or Selberg is

(II.1.2.1)

$$(4\pi)^{1-k-s}\Gamma(s') \sum \frac{a_n \bar{b}_n}{n^{s'}} = \int_{SL_2(\mathbb{Z})\backslash H} f(z)\bar{g}(z)E(z, s)y^{k-2}dxdy .$$

Here $s' = s + k - 1$, and $E(z,s)$ denotes the familiar Eisenstein series defined on the upper halfplane H by

$$E(z,s) = \sum_{\substack{(m,n)\in\mathbb{Z} \\ (m,n)=1}} \frac{y^s}{|mz+d|^{2s}} .$$

The idea is that the meromorphic continuation and functional equation of $\sum \frac{a_n \bar{b}_n}{n^s}$ can then be read off from the similar (known) properties of $E(z,s)$.

Given π_1, π_2 automorphic cuspidal representations of $GL_2(\mathbb{A})$, the analogue of the right-hand side of (II.1.2.1) is

$$\int_{Z_{\mathbb{A}} GL_2(F)\backslash GL_2(\mathbb{A})} \varphi_1(g)\varphi_2(g)E^f(g,s)dy = \mathcal{Z}(s,\varphi_1,\varphi_2,E^f)$$

where the φ_i are cusp forms realizing π_1 and π_2, and $E^f(g,s)$ is an appropriately defined Eisenstein series generalizing $E(z,s)$. More precisely, if χ denotes the *product* of the central characters of π_1 and π_2, then we set

$$E^f(g,s) = E(g, f_{s,\chi}) = \sum_{\gamma\in B_F\backslash G_F} f(\gamma g) ,$$

where $f = f_{s,\chi}$ belongs to the representation of $G_{\mathbf{A}} = GL_2(\mathbb{A})$ induced from the character $\begin{bmatrix} a & x \\ 0 & b \end{bmatrix} \longrightarrow |\frac{a}{b}|^{s-1/2}\chi^{-1}(b)$ of the Borel subgroup $B_{\mathbb{A}}$ of $G_{\mathbb{A}}$. We denote the space of this induced representation $Ind\ |\frac{a}{b}|^{s-1/2}\chi^{-1}(b)$ by $\mathbb{H}^{1,\chi^{-1}}(s)$. Plugging in Fourier expansions of the form (I.1.2.3) for φ_1 and φ_2, and using the definition just given for $E^f(g,s)$ (cf. formula (I.2.1.1), we obtain the product expansion

$$\mathcal{Z}(s,\varphi_1,\varphi_2,E^f) = \Pi\mathcal{Z}(s,W_v^1,W_v^2,f_v) \qquad \text{(II.1.2.2)}$$

where $W_v^i \in \mathcal{W}(\pi_{i,v},\psi_v)$, f_v is in $\mathbb{H}^{1,\chi_v^{-1}}(s)$ and

$$\mathcal{Z}(s,W_v^1,W_v^2,f_v) = \int_{ZN\backslash GL_2(F_v)} W_v^1(g)W_v^2(g)f_v(g)dg .$$

The first 120 pages of [Jacquet] are devoted to proving a functional equation for these local zeta-integrals and defining the (local) factors $L(s,\pi_1\times\pi_2)$ and $\varepsilon(s,\pi_1\times\pi_2,\psi)$. As is typical in this theory, the archimedean places – though in some ways simpler conceptually – often produce thornier computational problems. In any case, the end result is a complete Jacquet-Langlands type theory of L-functions for the group

$$G = GL(2) \times GL(2) \ ,$$

with $r : \ {}^LG \longrightarrow GL_4(\mathbb{C})$ the tensor product of the standard representations of $GL_2(\mathbb{C})$, i.e.,

$$r(g_1,g_2) = g_1 \otimes g_2 \quad \text{for} \quad (g_1,g_2) \in GL_2(\mathbb{C}) \times GL_2(\mathbb{C}) = {}^L G \ .$$

For future reference, it will be useful to analyze this work of Jacquet's in much more detail. However, so as not to unduly interrupt the flow of our chronological survey of L-functions, we shall limit this lengthy discussion to the appendix below, and advise the reader to study it carefully *only* in connection with Sections 1.3 and 1.5 of Chapter III.

APPENDIX TO SECTION (1.2):

Analysis and Reformulation of the Method of Rankin-Selberg-Jacquet for $GL(2) \times GL(2)$

The novel feature of Jacquet's approach is his ingenious use of Schwartz-Bruhat functions (and their Fourier transforms) on the affine space $\mathbb{A}^2$ in normalizing the Eisenstein series $E(g,s)$, and defining the local factors $L(s, \pi_1 \times \pi_2)$ as g.c.d.'s. As we shall see in §1 of Chapter III, this use of the Fourier transform has no straightforward generalization in the context of the Rankin-Selberg theories of [PSR1,2] and [GePS5]. Therefore, in a separate paragraph below, we shall sketch an alternate approach to Jacquet's theory which is based instead on the theory of intertwining operators (between $H^{1,\chi^{-1}}(s)$ and $H^{\chi^{-1},1}(1-s)$). This second approach does indeed seem to work for general Rankin-Selberg convolutions, as we shall eventually explain in Chapter III.

(A.1) We begin by recalling Jacquet's definition of $E(g,s,f)$. For this, we look upon the collection of spaces $\mathbb{H}^{1,\chi^{-1}}(s)$ as a fiber bundle over $\mathbb{C}$, with sections $f_s(g)$ such that

$$f_s\left(\begin{bmatrix} a & x \\ 0 & b \end{bmatrix} g\right) = \left|\frac{a}{b}\right|^s \chi^{-1}(b) f_s(g) \tag{A.1.1}$$

for all $\begin{bmatrix} a & x \\ 0 & b \end{bmatrix}$ in $B_{\mathbb{A}}$. Jacquet's idea is to consider $E(s,g,f)$ only for *special* sections of $\mathbb{H}^{1,\chi^{-1}}(s)$ such that $E(g,s,f)$ has only finitely many poles in $\mathbb{C}$, and fits in well with the local theory.

In more detail, we attach a section $s \longrightarrow f^{\Phi}_{s,\chi}(g)$ to each Schwartz-Bruhat function $\Phi = \Pi_v \Phi_v$ in $\mathcal{S}(\mathbb{A}^2)$ as follows. Set

$$f^{\Phi}_{s,\chi}(g) = |det\ g|^{s'+1/2} \int_{\mathbb{A}^x} \Phi([0,t]g)|t|^{2s'+1}\chi(t)d^x t\ , \tag{A.1.2}$$

where

$$s' = s - \frac{1}{2}\ .$$

It is easy to check that $f^{\Phi}_{s,\chi}$ belongs to $\mathbb{H}^{1,\chi^{-1}}(s)$ for each Φ in $\mathcal{S}(\mathbb{A}^2)$; indeed, this is a formal computation, valid at first only for $Re(s)$ large, but then for all s' thanks to the global theory of zeta-integrals ([Tate1]). We note that the substitution $s \longrightarrow 1-s$ corresponds to the substitution $s' \longrightarrow -s'$, since $s' = s - \frac{1}{2}$.

Using the sections $s \longrightarrow f^{\Phi}_{s,\chi}(g)$ Jacquet proves that $E(g, f^{\Phi}_{s,\chi})$, and hence also the global zeta-integral

$$\mathcal{Z}(s, \varphi_1, \varphi_2, f^{\Phi}_{s,\chi}) = \int_{Z_{\mathbb{A}} GL_2(A)\backslash GL_2(\mathbb{A})} \varphi_1(g)\varphi_2(g)E(g, s, f^{\Phi}_{s,\chi})dg \ ,$$

is actually entire in $\mathbb{C}$, except for possible poles at s_0 such that $||^{2-2s_0} = \chi$ or $||^{2s_0} = \chi^{-1}$. (We describe an alternate approach to the holomorphy of $E(g, f_s)$ in A.2 below.) Moreover, we have functional equations of the form

$$\text{(A.1.3)} \qquad E(g, f^{\Phi}_{s,\chi}) = \tilde{E}(g, f^{\hat{\Phi}}_{1-s,\chi^{-1}}\chi^{-1}(det\ g))$$

and

$$\text{(A.1.4)} \qquad \mathcal{Z}(s, \varphi_1, \varphi_2, f^{\Phi}_{s,\chi}) = \tilde{\mathcal{Z}}(1-s, \varphi_1, \varphi_2, f^{\hat{\Phi}}_{1-s,\chi^{-1}})$$

where $\hat{\Phi}$ denotes the Fourier transform

$$\hat{\Phi}(x, y) = \int \Phi(u, v)\psi(yu - xv)dudv \ .$$

The exact nature of this "contragredient" function $\tilde{\mathcal{Z}}$ will be explained locally after (A.1.8) below. The Euler product expansion (II.1.2.2), taken together with the analytic properties of $\mathcal{Z}(s, \varphi_1, \varphi_2, f^{\Phi}_{s,\chi})$ just described, therefore completes Steps 1 and 2 of the L-function machine sketched in (I.1.7). It remains "only" to treat the *local* zeta-integrals $\mathcal{Z}(s, W_1, W_2, f^{\Phi}_{s,\chi})$.

The first thing to prove is that each of these local zeta-integrals defines a meromorphic function of s, in fact, a rational function in q^{-s} when the local field F is nonarchimedean (with residue characteristic q). This results immediately from the asymptotic properties of the Whittaker functions

$W_i(g)$ (cf. I.(1.3.2)), their right invariance by some open subgroup, and the Iwasawa decomposition for $GL(2)$ which implies

$$\mathcal{Z}(s,W_1,W_2,f_s^{\Phi}) = \sum_i f_s(k_i)\int W_1\left(\begin{bmatrix} a & 0\\ 0 & 1\end{bmatrix} k_i\right) W_2\left(\begin{bmatrix} a & 0\\ 0 & 1\end{bmatrix} k_i\right)|a|^s d^x a\,. \tag{A.1.5}$$

More precisely, the family of functions $\mathcal{Z}(s,W_1,W_2,f_s^{\Phi})$, for arbitrary W_1,W_2 and Φ, generates a fractional ideal $\mathcal{U}_p$ of the ring $\mathbb{C}[q^{-s},q^s]$ inside $\mathbb{C}(q^{-s})$. So proceeding exactly as in Jacquet-Langlands theory (cf. I.(1.6)), we may introduce "the" Euler factor $L(s,\pi_1\times\pi_2)=Q_0(q^{-s})^{-1}$ which generates $\mathcal{U}_p$. The fact that $\mathcal{Z}(s,W_1,W_2,\Phi)$ may be chosen equal to 1 once again, plays a crucial role in the conclusion that $L(s,\pi_1\times\pi_2)$ may be taken equal to 1 over a polynomial $Q_0(q^{-s})$ with $Q_0(0)=1$.

We recall that $f^{\Phi}_{\chi,s}$, being the local equivalent of (A.1.2), is defined by the formula

$$f^{\Phi}_{s,\chi}(g)=|det\ g|_F^{s'+1/2}\int_{F^x}\Phi([0,t]g)|t|^{2s'+1}\chi(t)d^x t \tag{A.1.6}$$

with $\Phi\in\mathcal{S}(F^2)$. Using the theory of the local Whittaker models $\mathcal{W}(\pi_i,\psi)$, Jacquet establishes the functional equation

$$\tilde{\mathcal{Z}}(1-s,W_1,W_2,f^{\hat{\Phi}}_{1-s,\chi^{-1}})=\gamma(s,\pi_1\times\pi_2)\mathcal{Z}(s,W_1,W_2,f^{\Phi}_{\chi,s})\,, \tag{A.1.7}$$

where

$$\tilde{\mathcal{Z}}(s,W_1,W_2,f^{\Phi}_{s,\chi})=\int_{NZ\backslash G}W_1(g)W_2(g)f^{\Phi}_{s,\chi}(g)\chi^{-1}(det\ g)dg\,. \tag{A.1.8}$$

Since this last zeta-integral is simply the "contragredient" version of $\mathcal{Z}(s,W_1,W_2,f)$ (cf. the remarks in Section I.1.5 for $GL(2)$), we let $L(s,\tilde{\pi}_1\times\tilde{\pi}_2)$ denote the Euler factor generating the fractional ideal $\{\tilde{\mathcal{Z}}(s,W_1,W_2,f^{\Phi}_{s,\chi})\}$. Then, just as in Section I.1.6, we rewrite (A.1.7) in the form

$$\frac{\tilde{\mathcal{Z}}(1-s,W_1,W_2,f^{\hat{\Phi}}_{1-s,\chi^{-1}})}{L(1-s,\tilde{\pi}_1\times\tilde{\pi}_2)}=\varepsilon(s,\pi_1\times\pi_2)\frac{\mathcal{Z}(s,W_1,W_2,f^{\Phi})}{L(s,\pi_1\times\pi_2)}$$

where

$$\varepsilon(s, \pi_1 \times \pi_2) = \frac{\gamma(s, \pi_1 \times \pi_2) L(s, \pi_1 \times \pi_2)}{L(1-s, \tilde{\pi}_1 \times \tilde{\pi}_2)}$$

is a "monomial" of the form cq^{-sb}. Analogous to the GL_2 theory, the functional equation results from the uniqueness of a certain invariant form, in this case a *trilinear* form on $\mathcal{W}(\pi_1, \psi) \times \mathcal{W}(\pi_2, \psi) \times \mathbb{H}^{1,\chi^{-1}}(s)$.

Remark. The map $\Phi \longrightarrow f^{\Phi}_{s,\chi}$ intertwines the natural GL_2-module $\mathcal{S}(F^2)$ with an *invariant* subspace of $\mathbb{H}^{1,\chi^{-1}}(s)$. But $\mathbb{H}^{1,\chi^{-1}}(s)$ is irreducible unless $\chi(x) = |x|^{\pm 1 - 2s'}$ (cf. [JL], p.99). It follows that the space of functions $\{f^{\Phi}_{s,\chi}\}$ (at least generically) exhausts $\mathbb{H}^{1,\chi^{-1}}(s)$, as Φ runs through $\mathcal{S}(F^2)$.

We complete our discussion of the local theory by referring to the unramified computation

$$\mathcal{Z}(s, W_1, W_2, f^{\Phi}_{s,\chi}) = L(2s, \chi) \int W_1 \begin{pmatrix} a & 0 \\ 0 & 1 \end{pmatrix} W_2 \begin{pmatrix} a & 0 \\ 0 & 1 \end{pmatrix} |a|^{s-1} d^x a$$

$$= L(2s, \chi) \left\{ \frac{\prod_{i,j=1}^{2} L(s, \mu_i \nu_j)}{L(2s, \chi)} \right\} = \frac{1}{det(I - t_{\pi_1} \otimes t_{\pi_2} q^{-s})}$$

(cf. [Jacquet], pp. 32, 33). Here π_1 and π_2 are the unramified representations corresponding to t_{π_1} and t_{π_2} in $GL_2(\mathbb{C})$, W_1 and W_2 are the unique $GL_2(O_v)$ fixed vectors in $\mathcal{W}(\pi_1, \psi)$ which equal 1 at e, Φ is the characteristic function of $O_v \times O_v$, and χ is unramified; note that (from (A.1.6)) we indeed have

$$f^{\Phi}_{s,\chi}(k) = f^{\Phi}_{s,\chi}(e) = L(2s, \chi) = \frac{1}{1 - \chi(\tilde{\omega}_v) q_v^{-2s}}$$

for all k in $GL_2(O_v)$.

Thus we see that the sections $f^{\Phi}_{s,\chi}$ function perfectly (in conjunction with the Fourier transform): globally the resulting Eisenstein series are easily shown to have only finite many poles, locally the zeta-integrals give the correct Euler-Langlands factors $L(s, \pi_1 \times \pi_2)$, and both (the local and global) functional equations involve the compatible substitution $\Phi \longrightarrow \hat{\Phi}$.

In a sense which will become clearer in the next paragraph, Jacquet's choice of sections $s \longrightarrow f^{\Phi}_{s,\chi}(g)$ already has built into it the "normalizing factor" $L(2s,\chi)$ which globally knocks out (infinitely many) unwanted poles of $E(g,s)$, and locally produces the correct Langlands-Euler factor.

(A.2) An Alternate Approach to [Jacquet]

Our purpose is to redo Jacquet's theory without recourse to Fourier transforms or the use of Schwartz-Bruhat functions in defining permissible sections $s \longrightarrow f_s(g)$.

We begin with the Eisenstein series

$$E(s,g,f_s) = \sum_{B_F\backslash G_F} f_s(\gamma g) \ ,$$

where $f_s \in \mathbb{H}^{1,\chi^{-1}}(s) = Ind_{B_{\mathbb{A}}}^{G_{\mathbb{A}}} |\frac{a}{b}|^{s-1/2}\chi^{-1}(b)$. By the general theory of Eisenstein series sketched in Section I.2, we know that the analytic properties of $E(s,g,f_s)$ are bound up with those of the operator

$$\text{(A.2.1)} \qquad f \longrightarrow M(s)f(g) = \int_{\mathbb{A}} f\left(w\begin{pmatrix}1 & x\\ 0 & 1\end{pmatrix} g\right) dx$$

intertwining $\mathbb{H}^{1,\chi^{-1}}(s)$ with $\mathbb{H}^{\chi^{-1},1}(1-s)$. In the present context, relation (I.2.5.3) becomes an identity of the form

$$\text{(A.2.2)} \qquad M(s)f = \left(\prod_{v\in S} A(s,\chi_v)f_v\right) \frac{L_S(2s',\chi)}{L_S(2s'+1,\chi)} \prod_{v\notin S} \tilde{f}_v$$

where S is a finite set of places of F (including the infinite places) outside of which χ_v is unramified, and $f = \Pi f_v$ with f_v a $GL_2(O_v)$-fixed (spherical) function in $\mathbb{H}^{1,\chi^{-1}}(s)$ for each $v \notin S$. Recall that $s' = s - \frac{1}{2}$, and note that the discrepancy between (A.2.2) and (I.2.5.3), i.e., between s and $2s$, comes from the fact that we are working with GL_2 in place of SL_2; cf. also Remark I.2.5.4 (ii).

Our first task is to prove that $M(s)f$ (and hence $E(g,s,f)$) has only finitely many poles in $\mathbb{C}$ (coming from those of $L(2s',\chi)$) *provided* $f =$

Πf_v is an appropriately well chosen section for $\mathbb{H}^{1,\chi^{-1}}(s)$. In particular, let us consider only functions $f = \Pi f_v$ such that for *almost all* f_v,

(i) $f_v = L(2s'+1, \chi_p) f_v^0$, with f_v^0 the unique $GL_2(O_v)$ fixed function in $\mathbb{H}^{1,\chi^{-1}}(s)$ which is identically 1 on $GL_2(O_v)$; and *at a finite number of places*, either:

(ii) f_v *restricted* to $GL_2(O_v)$ is independent of s; or

(iii) f_v is of the form $A^*(1-s, \chi_v^{-1}) g_v$, where $A^*(1-s, \chi_v^{-1})$ is the (suitably normalized) intertwining operator taking $\mathbb{H}^{\chi^{-1},1}(1-s)$ to $\mathbb{H}^{1,\chi^{-1}}(s)$, and g_v in $\mathbb{H}^{\chi^{-1},1}(1-s)$ is of type (ii), i.e., its restriction to $GL_2(O_v)$ is independent of s.

The need for normalizing $A(s)$ (or $A(1-s)$), as well as the need for including possibility (iii) at all, will be clear from the form of the functional equation for $\mathcal{Z}(s, W_1, W_2, f)$ discussed after the Proposition below. For the moment, let us be content to simply describe $A^*(s)$. By $\gamma(s, \chi_v)$ we denote the abelian gamma factor

$$\gamma(s, \chi_v) = \frac{\varepsilon(s, \chi_v) L(1-s, \chi_v^{-1})}{L(s, \chi_v)}$$

which arises in Tate's local theory of L-functions for $GL(1)$ (cf. Sections I.1.3 and 1.7); then

$$A^*(s, \chi) = \gamma(2s', \chi) A(s, \chi) \ .$$

The immediate significance of this normalization is that A^* (as opposed to $A(s)$) satisfies the simple functional equation

$$A^*(s, \chi_v) A^*(1-s, \chi_v^{-1}) = I \ .$$

For a discussion of such normalizations in the general context of reductive G, see [Keys-Sha].

Proposition (A.2.3). With $f = \Pi f_v$ as above, $E(g, s, f)$ (or $M(s) f$) is meromorphic with only finite many poles in $\mathbb{C}$, namely those of $L(2s', \chi)$.

Proof. Consider the expression for $M(s)f$ given by (A.2.2). Then we have

$$M(s)f = \left(\prod_{v\in S} A(s,\chi_v)f_v\right) L_S(2s',\chi_v) \prod_{v\notin S} \tilde{f}_v^0$$

or

$$M(s)f = \left(\prod_{v\in S} \frac{1}{L(2s',\chi_v)} A(s,\chi_v)f_v\right)\left(\prod_{v\notin S} \tilde{f}_v^0\right) L(2s',\chi) .$$

In order to establish the Proposition, it therefore suffices to prove the following Lemma.

Lemma A.2.4. For each v in S, and f_v as above,

$$\frac{1}{L(2s',\chi_v)} A(s,\chi_v)f_v(g)$$

is *entire* in s.

Proof. If f_v is of type (i), then $\frac{A(s,\chi_v)}{L(2s',\chi_v)} f_v = \frac{L(2s',\chi_v)L(2s'+1,\chi_v)}{L(2s'+1,\chi_v)L(2s',\chi_v)} \tilde{f}_v^0 = \tilde{f}_v^0$. I.e., $\frac{A(s,\chi_v)}{L(2s',\chi_v)} f_v$ is clearly entire in s, and we may now assume f_v is of type (ii) or (iii).

By definition,

$$A(s,\chi_v)f_v(g) = \int_{F_v} f_{v,s}(w\begin{pmatrix}1 & x\\ 0 & 1\end{pmatrix}g)dx$$

$$= \int_{|x|\text{ small}} f_{v,s}(w\begin{pmatrix}1 & x\\ 0 & 1\end{pmatrix}g)dx + \int_{|x|\text{ large}} f_{v,s}(w\begin{pmatrix}1 & x\\ 0 & 1\end{pmatrix}g)dx .$$

Since this first integral converges absolutely for all s, the meromorphic behavior of $A(s,\chi_v)f_v(s)$ depends only on the second integral. Note that the property of f_v being of type (i), (ii) or (iii) is preserved under right translations by k in $GL_2(O_v)$, whereas b in B simply moves (across $w\begin{pmatrix}1 & x\\ 0 & 1\end{pmatrix}$ and) out of the integral in question. Therefore we may assume $g = bk$ with $b = e = k$.

For simplicity, let us suppose v is a finite place. Since f_v is then right invariant by $\begin{bmatrix}1 & 0\\ -x^{-1} & 1\end{bmatrix}$ *for* $|x|$ *large*, the Iwasawa decomposition

$$w\begin{bmatrix}1 & x\\ 0 & 1\end{bmatrix}\begin{bmatrix}1 & 0\\ -x^{-1} & 1\end{bmatrix} = \begin{bmatrix}x^{-1} & -1\\ 0 & x\end{bmatrix}$$

implies that the second integral above (with $g = e$) equals

$$f_v(e)\int_{|x| \text{ large}} \chi_v^{-1}(x)|x|^{-2s}dx \ .$$

Thus we need only examine the analytic nature of the expression

$$\frac{f_{v,s}(e)}{L(2s',\chi_v)}\int_{|x| \text{ small}} \chi_v(x)|x|^{2s'-1}dx \ .$$

If f_v is of type (ii) above, then $f_{v,s}(e)$ is independent of s, and $\frac{A(s,\chi_v)f_v}{L(2s',\chi_v)}$ is an entire function of s, whether or not χ_v is unramified (since the integral $\int_{|x| \text{ small}} \chi_v(x)|x|^{2s'-1}dx$ has the same meromorphic behavior as $L(2s',\chi_v)$). On the other hand, if f_v is of type (iii), then $f_{v,s}(g) = A^*(1-s,\chi_v^{-1})g_{v,s}$, with $g_{v,s}$ of type (ii). Then since $A^*(s)A^*(1-s) = I$, we have

$$\begin{aligned}\frac{1}{L(2s',\chi_v)}A(s,\chi_v)f_{v,s} &= \frac{A^*(s,\chi_v)f_{v,s}}{L(2s',\chi_v)\gamma(2s',\chi_v)}\\ &= \frac{A^*(s)A^*(1-s)g_{v,s}}{\varepsilon(2s',\chi_v)L(1-2s',\chi_v^{-1})} = \frac{1}{\varepsilon(2s',\chi_v)L(1-2s',\chi_v^{-1})}g_{v,s} \ .\end{aligned}$$

But $g_{v,s}$ is, by assumption, entire in s, and so clearly is $1/\varepsilon(2s',\chi_v)L(1-2s',\chi_v^{-1})$. This completes the proof of the Lemma and the Proposition.

Notice that we have "normalized" our sections $f = \Pi f_v$ by the factor $L_S(2s'+1,\chi)$ precisely so that the troublesome denominator in (A.2.2) can not contribute (by way of its zeros) infinitely many poles to $M(s)f$ (and hence $E(g,s)$). It is interesting to compare this technique of normalization with Jacquet's more sophisticated (but more special) approach. In the latter approach, the fact that $f^{\Phi}_{s,\chi}(k) \equiv L(2s'+1,\chi)$ means that these poles are automatically cancelled, i.e., the normalizing factor is "built into" the choice of section $f^{\Phi}_{s,\chi}(g)$.

We turn next to the functional equations for $\mathcal{Z}(s,\varphi_1,\varphi_2,E)$ and $\mathcal{Z}(s,W_1,W_2,f)$. As already remarked, the general theory of Eisenstein series tells us that

$$E(g,s,f_s) = E(g,1-s,M(s)f) \ ,$$

whence the global functional equation

$$\mathcal{Z}(s, \varphi_1, \varphi_2, E^f) = \mathcal{Z}(1-s, \varphi_1, \varphi_2, E^{M(s)f}) .$$

For the local zeta-integrals, it can be shown that

$$\mathcal{Z}(1-s, W_1, W_2, A(s)f) = \Gamma(s, \pi_1 \times \pi_2)\mathcal{Z}(s, W_1, W_1, f) , \tag{A.2.5}$$

with $\Gamma(s)$ meromorphic. (In fact, $\Gamma(s)$ is rational in q_v^{-s} for F non-archimedean.) The connection between these results and those of Jacquet (cf. (A.1.3) and (A.1.7)) may now be explained as follows.

Consider the "local" diagram

$$\begin{array}{ccc} S(F^2) \ni \Phi & \xrightarrow{\text{Fourier transform}} & \hat{\Phi} \\ \downarrow & & \downarrow \; \hat{\Phi} \to \tilde{f}(g) = f^{\hat{\Phi}}_{1-s,\chi^{-1}}(g)\frac{1}{\chi}(det\ g) \\ \mathbb{H}^{1,\chi^{-1}}(s) \ni f & \xrightarrow{A^*(s,\chi)} & \tilde{f} \in \mathbb{H}^{\chi^{-1},1}(1-s) \end{array} \tag{A.2.6}$$

where the left vertical map is $\Phi \to f(g) = f^{\Phi}_{s,\chi}(g)$, and the bottom horizontal map is simply our normalized operator $A^*(s)$, i.e. the operator $A(s)$ multiplied by the "abelian gamma factor"

$$\gamma(2s', \chi) = \frac{\varepsilon(2s', \chi)L(1-2s', \chi^{-1})}{L(2s', \chi)} . \tag{A.2.7}$$

It is a (wonderful) fact that this diagram commutes (cf. Lemma 14.7.1 of [Jacquet]), *and* that the product of these local factors γ (over all the places of F) equals 1. (This last fact is equivalent to the functional equation for $L(s, \chi)$; cf. [Tate1]. Our functional equations above therefore reduce precisely to those of Jacquet when $f = f^{\Phi}$. Indeed, in the local equation we set $A^*(s) = \gamma(2s, \chi)A(s)$ so that the equation reads

$$\mathcal{Z}(1-s, W_1, W_2, A^*(s)f) = \gamma(s, \pi_1 \times \pi_2)\mathcal{Z}(s, W_1, W_2, f) \tag{A.2.8}$$

with $\gamma(s, \pi_1 \times \pi_2) = \gamma(2s', \chi)\Gamma(s, \pi_1 \times \pi_2)$; in the global case the gamma factor simply disappears.) It remains to describe the local L and ε factors $L(s, \pi_1 \times \pi_2)$ and $\varepsilon(s, \pi_1 \times \pi_2)$ in terms of the family of zeta-integrals

$\mathcal{Z}(s, W_1, W_2, f_s)$, as was done for $GL(2)$ in Sections I.(1.3) and (1.6). The crucial point is that this family of integrals generates a fractional ideal $\mathcal{U}_v$ in $\mathbb{C}[q^s, q^{-s}]$, as f_s runs through "good sections" of type (i), (ii) or (iii) above. The proof that there exists a common denominator for the integrals $\mathcal{Z}(s, W_1, W_2, f_s)$, say for f_s of type (ii), proceeds exactly as in Jacquet's theory. Indeed, recall

$$\mathcal{Z}(s, W_1, W_2, f_s) = \sum_i f_s(k_i) \int W_1(\begin{bmatrix} a & 0 \\ 0 & 1 \end{bmatrix} k_i) W_2(\begin{bmatrix} a & 0 \\ 0 & 1 \end{bmatrix} k_i) |a|^s d^x a \ .$$

Since $f_s(k_i)$ equals $L(2s'+1, \chi)$, or is independent of s, it is clear by the usual theory of asymptotics for $W_i[\begin{pmatrix} a & 0 \\ 0 & 1 \end{pmatrix}]$ that the functions $\mathcal{Z}(s, W_1, W_2, f_s)$ admit a g.c.d. On the other hand, for f_s of type (iii), the local functional equation (A.2.8) reduces matters again to the case of type (ii).

To see that $\mathcal{Z}(s, W_1, W_2, f_s) = 1$ for appropriately chosen W_i and f, we fix f_s to be of type (ii) and such that its restriction to $GL_2(O_v)$ has support in a compact subgroup fixing W_1 and W_2, with W_i the characteristic function of $(O_v)^x$.

As usual, we let $L(s, \pi_1 \times \pi_2)$ denote the g.c.d. of the integrals $\mathcal{Z}(s, W_1, W_2, f_s)$ for all "good sections" f_s (of type (i), (ii) or (iii)). We also let $L(1-s, \tilde{\pi}_1 \times \tilde{\pi}_2)$ denote the corresponding generator of the ideal $\mathcal{Z}(1-s, W_1, W_2, A^*(s)f)$. By definition, there exists a finite number of W_1^i, W_2^i and f_s^i such that

$$\sum_{i=1}^{N} \mathcal{Z}(s, W_1^i, W_2^i, f_v^i) = L(s, \pi_1 \times \pi_2)$$

and similarly for $L(s, \tilde{\pi}_1 \times \tilde{\pi}_2)$. In particular, when all data in sight is unramified, we have

$$\mathcal{Z}(s, W_1, W_2, f) = f(e) \left(\int W_1 \begin{pmatrix} a & 0 \\ 0 & 1 \end{pmatrix} W_2 \begin{pmatrix} a & 0 \\ 0 & 1 \end{pmatrix} |a|^{s-1} d^x a \right)$$

with

$$f(e) = L(2s, \chi) \ .$$

Thus we obtain the same local Langlands factor $L(s, \pi_1 \times \pi_2)$ computed in [Jacquet] (but only after forcibly bringing into play the "normalizing factor" $L(2s, \chi)$).

In order to get the best possible global functional equation for $L(s, \pi_1 \times \pi_2)$, we need to analyze the local functional equation

$$\frac{\mathcal{Z}(1-s, W_1, W_2, A^*(s)f)}{L(1-s, \tilde{\pi}_1 \times \tilde{\pi}_2)} = \varepsilon(s, \pi_1 \times \pi_2) \frac{\mathcal{Z}(s, W_1, W_2, f)}{L(s, \pi_1 \times \pi_2)} \tag{A.2.9}$$

where $A^*(s) = \gamma(2s', \chi)A(s)$ is our "normalized" intertwining operator, and

$$\varepsilon(s, \pi_1 \times \pi_2) = \gamma(s, \pi_1 \times \pi_2) L(s, \pi_1 \times \pi_2) L(1-s, \tilde{\pi}_1 \times \tilde{\pi}_2) .$$

As we have already remarked (and as follows alternatively from the commutativity of the diagram (A.2.6)),

$$A^*(s)A^*(1-s) = I . \tag{A.2.10}$$

The significance of (A.2.10) at this juncture is that it allows us to apply the above functional equation twice, and conclude (just as in I.(1.6) for $GL(2)$) that

$$\gamma(s, \pi_1 \times \pi_2)\gamma(1-s, \tilde{\pi}_1 \times \tilde{\pi}_2) = 1 \tag{A.2.11}$$

and similarly

$$\varepsilon(s, \pi_1 \times \pi_2)\varepsilon(1-s, \tilde{\pi}_1 \times \tilde{\pi}_2) = 1 . \tag{A.2.12}$$

As before, this implies $\varepsilon(s, \pi_1 \times \pi_2)$ has neither zeros nor poles in $\mathbb{C}$.

Finally, we use these facts to establish the analytic continuation and functional equation of $L(s, \pi_1 \times \pi_2) = \Pi L_v(s, \pi_1 \times \pi_2)$. For each place v, we may pick functions f_v such that

$$\Pi \mathcal{Z}(s, W_1, W_2, f_v) = L(s, \pi_1 \times \pi_2) .$$

(More precisely, we need to work with a finite sum of such products on the left hand side, but this won't change the validity of our arguments.) Then

$$L(s,\pi_1\times\pi_2)=\mathcal{Z}(s,\varphi_1,\varphi_2,E_f)=\mathcal{Z}(1-s,\varphi_1,\varphi_2,E_{M(s)f})$$

$$=\left(\prod_{p\in S}\mathcal{Z}(1-s,W_1,W_2,A^*(s)f_v)\right)L_S(1-s,\tilde{\pi}_1\times\tilde{\pi}_2)$$

$$=\left(\prod_{p\in S}\frac{\mathcal{Z}_v(1-s,W_1,W_2,A^*(s)f_v)}{L_v(1-s,\tilde{\pi}_1\times\tilde{\pi}_2)}\right)L(1-s,\tilde{\pi}_1\times\tilde{\pi}_2)\ .$$

But by the local functional equation, the last expression in parentheses equals $(\Pi_{p\in S}\varepsilon_v(s,\pi_1\times\pi_2))$ multiplied by $(\Pi_{p\in S}\frac{\mathcal{Z}(s,W_1,W_2,f)}{L(s,\pi_1\times\pi_2)})$ (which equals 1 by our choice of f_v's). Thus we get the desired functional equation

$$L(s,\pi_1\times\pi_2)=\varepsilon(s,\pi_1\times\pi_2)L(1-s,\tilde{\pi}_1\times\tilde{\pi}_2)$$

with "monomial" factor ε.

The analytic continuation of $L(s,\pi_1\times\pi_2)$, on the other hand, follows from the identity

$$L(s,\pi_1\times\pi_2)=\int\varphi_1(g)\varphi_2(g)E(g,s,f_s)dg\ .$$

In particular, according to what we just learned about the poles of $E(g,s,f_s)$ (for these good $f=\Pi f_v$), we know that the only possible poles of $L(s,\pi_1\times\pi_2)$ are those of $L(2s',\chi)$. This result is consistent with Jacquet's assertion that the only possible poles occur when $|\ |^{2-2s}=\chi$ or $|\ |^{2s}=\chi^{-1}$. When χ is not of the form $|\ |^{\sigma}_{\mathbb{A}}$, $L(s,\pi_1\times\pi_2)$ must be entire; on the other hand, if χ is of the form $|\ |^{\sigma}_{\mathbb{A}}$, then $L(s,\pi_1\times\pi_2)$ can have a pole only if the residue $\int\varphi_1(g)\varphi_2(g)|det\ g|^s dg$ is non-zero, i.e., $\pi_2\approx\tilde{\pi}_1$ modulo a power of $|det\ g|^s$. Generalizations of these results to $GL(n)\times GL(m)$ are the subject matter of Section 1.7 below.

1.3. Shimura's method.

Some important variations of the Rankin-Selberg method were introduced by Shimura in a series of works in the early 1970's. In [Shim1], an attempt is made to develop a theory of L-functions for modular forms of half-integral weight analogous to Hecke's theory for forms of integral weight. The starting point is an analogue of the basic identity of Rankin-Selberg which takes the following form:

$$\text{(II.1.3.1)} \quad \begin{aligned} &2(4\pi)^{-s}\Gamma(s)\left(\sum_{n=1}^{\infty}\Psi(n)a(n^2)n^{\nu-2s}\right)L(2s-\lambda-\nu+1,\Psi) \\ &= \int_{\mathcal{F}} f(z)\theta_{\Psi}(z)E(z,s)y^{s-1}dxdy\ . \end{aligned}$$

Here the coefficients $a(n^2)$ arise as Fourier coefficients of the modular form

$$f(z)=\sum a(n)e^{2\pi inz}$$

which is of half-integral weight $k/2$ with respect to some subgroup Γ of $SL_2(\mathbb{Z})$, $\Psi(n)$ defines some Dirichlet character of the integers, ν is 0 or 1 depending on whether $\Psi(-1)=1$ or -1, $L(s,\Psi)=\sum\Psi(n)n^{-s}$, $\theta_{\Psi}(z)$ is the theta function

$$\text{(II.1.3.2)} \qquad \theta_{\psi}(z)=\sum_{n=-\infty}^{\infty}\Psi(n)n^{\nu}e^{2\pi in^2z}$$

(itself a modular form of weight $\frac{2\nu+1}{2}$), $E(z,s)$ is again a familiar Eisenstein series on H, and $\mathcal{F}$ is a fundamental domain for $\Gamma\backslash H$. The point is that the well-known analytic properties of $E(z,s)$ may once again be used to read off similar properties for an automorphic L-function.

In this case, it is important to remark that $\sum\frac{a(n^2)}{n^s}$ is not exactly the natural L-function to attach to $f(z)$. The "Mellin transform" $\sum\frac{a(n)}{n^s}$ would be more natural, if not for the fact that it fails (in general) to have an Euler product expansion; this is exactly why Shimura treats $\sum\frac{a(n^2)}{n^s}$ in its place.

In [Shim2], yet another intriguing variation of the Rankin-Selberg identity is introduced, this time in order to extend Hecke's theory of L-functions for forms of *integral* weight to "symmetric squares" of these L-functions. More precisely, suppose once again that

$$f(z) = \sum_{n=1}^{\infty} a(n) e^{2\pi i n z}$$

is a cusp form of (*integral*) weight k for $SL_2(\mathbb{Z})$ and an eigenfunction of all the Hecke operators. As already remarked before, this means that

$$\begin{aligned} L_s(s,f) = \sum \frac{a(n)}{n^s} &= \prod_p (1 - a(p)p^{-s} + p^{k-1-2s})^{-1} \\ &= \prod_p \frac{1}{(1-\alpha_p p^{-s})(1-\beta_p p^{-s})} \,. \end{aligned}$$

By the *symmetric square* of this L-function we mean the Euler product

$$L^2(s,f) = \prod_p \frac{1}{(1-\alpha_p^2 p^{-s})(1-\alpha_p\beta_p p^{-s})(1-\beta_p^2 p^{-s})} \,. \tag{II.1.3.3}$$

The goal of [Shim2] is to establish the analytic continuation and functional equation of $L^2(s,f)$. This is accomplished by means of the basic identity

$$\begin{aligned} &2\pi^{k-1-s}(4\pi)^{-s}\Gamma(s)\Gamma(s-k+1)\zeta(s-k+1)L^2(s,f) \\ &= \int_{\Gamma\backslash H} f(z)\theta(z)\tilde{E}(z,s)\frac{dxdy}{y^2} \end{aligned} \tag{II.1.3.4}$$

where $\theta(z)$ is again the basic theta-function $\sum_{n=-\infty}^{\infty} e^{2\pi i n^2 z}$, but $\tilde{E}(z,s)$ is now an Eisenstein series of *half*-integral weight. Most of the difficulty in [Shim 2] arises from the need to analyze from scratch this not-so familiar type of Eisenstein series. The *method* of studying $L^2(s,f)$, however, is completely analogous to the method of Rankin-Selberg.

Henceforth, we shall refer to both these variations of the method of Rankin-Selberg sketched above as "Shimura's method". A Jacquet-Langlands type theory of L-functions attached to modular forms of half-integral weight, using a representation theoretic version of Shimura's method as well as Hecke's Mellin transforms, was subsequently developed in [GePS2-4]. On the other hand, Shimura's theory of the symmetric square was developed representation theoretically in [GeJa]. These and related matters will be discussed in Sections (1.5) and (1.6) below.

1.4. Hecke theory for $GL(n)$.

Recall (from Section 1.1) that by [GoJa] we know how to attach to each unitary irreducible representation $\pi = \otimes\pi_v$ of $GL_n(\mathbb{A}_F)$ an infinite Euler product

$$L(s,\pi) = \prod_v L(s,\pi_v)$$

where the local factors $L(s,\pi_v)$ (and $\varepsilon(s,\pi_v,\psi_v)$) are defined in terms of the local representations π_v. Moreover, if π is also automorphic cuspidal, we know that $L(s,\pi)$ extends to an entire function of s such that $L(s,\pi) = \varepsilon(s,\pi)L(1-s,\tilde{\pi})$, with $\varepsilon(s,\pi) = \Pi\varepsilon(s,\pi_v,\psi_v)$.

It is natural to ask now whether the *converse theorem* is true. If π is a unitary irreducible representation of $GL_n(\mathbb{A}_F)$, and *if* $L(s,\pi)$ satisfies the above properties, does it necessarily follow that π is automorphic cuspidal? Except in the most simple cases, these assumptions are *not* sufficient. For $n = 2$, Weil was the first to propose the correct strengthening of assumptions: "nice" properties of the L-function must be met not only by π, but also by all the representations $\pi\otimes\chi$, where χ is any automorphic representation of $GL(1)$ i.e., a character of the idele class group $F^x\backslash\mathbb{A}_F^x$. (Recall that by $\chi(g)$ we mean $\chi(det\ g)$.) Under these strengthened assumptions, the representation π is indeed automorphic cuspidal on $GL(2)$; cf. [Wel] and [JL].

Much of the work of Jacquet, Piatetski-Shapiro and Shalika in the 1970's amounts to an attempt to prove an analogous "converse theorem" in the context of $GL(n)$, $n \geq 3$. It turns out that for $n = 3$ the result holds just as for $GL(2)$: if $L(s,\pi\otimes\chi)$ is "nice" for all twists $\pi\otimes\chi$, then π is automorphic cuspidal.

This result was first discovered by Piatetski-Shapiro ([PS1]), with the complete theory appearing in [JPSS1]. On the other hand, for $n \geq 4$ it is necessary to "twist" $L(s,\pi)$ by all automorphic cuspidal representations π' of the groups $GL(j)$, $1 \leq j \leq n-2$; this result, in turn, depends on the corresponding theory of "Rankin convolutions" $L(s, \pi \times \pi')$ (to be reviewed in Section (1.7) below).

To prove the converse theorem for $GL(3)$, it is necessary to extend the method of Hecke-Jacquet-Langlands from $GL(2)$ to $GL(3)$. This gives a direct proof of the analytic properties of $L(s,\pi)$ which can be "inverted" (unlike the proof of Godement-Jacquet). Globally, the zeta-function introduced is

$$\mathcal{Z}(s,\varphi) = \int_{F^x\backslash\mathbb{A}^x} V\left(\begin{bmatrix} a & 0 & 0 \\ 0 & 1 & 0 \\ 0 & 0 & 1 \end{bmatrix}\right) |a|^{s-1} d^x a$$

where φ is a cusp form in the space of $\pi = \otimes \pi_v$, and

$$V\left(\begin{bmatrix} a & 0 & 0 \\ 0 & 1 & 0 \\ 0 & 0 & 1 \end{bmatrix}\right) = \iint_{(F\backslash\mathbb{A})^2} \varphi\left(\begin{bmatrix} 1 & 0 & u \\ 0 & 1 & v \\ 0 & 0 & 1 \end{bmatrix}\begin{bmatrix} a & 0 & 0 \\ 0 & 1 & 0 \\ 0 & 0 & 1 \end{bmatrix}\right) \psi(-v)du dv \,.$$

Viewing $\left\{\begin{bmatrix} 1 & 0 & u \\ 0 & 1 & v \\ 0 & 0 & 1 \end{bmatrix}\right\} = U^P$ as the unipotent radical of the standard parabolic subgroup of $GL(3)$ of type (2,1), we may view $V\left(\begin{bmatrix} a & & \\ & 1 & \\ & & 1 \end{bmatrix}\right)$ as the "ψ-th Fourier coefficient" of φ along U^P, and $\mathcal{Z}(s,\varphi)$ as a kind of *Mellin transform* (just as $\mathcal{Z}(s,\varphi)$ is directly the Mellin transform of φ in case $n=2$). The basic identity for $\mathcal{Z}(s,\varphi)$ then takes the form

$$\mathcal{Z}(s,\varphi) = \int_{\mathbb{A}^x} W_\varphi^\theta\left(\begin{bmatrix} a & 0 & 0 \\ 0 & 1 & 0 \\ 0 & 0 & 1 \end{bmatrix}\right) |a|^{s-1} d^x a$$

where $\theta_N\left(\begin{bmatrix} 1 & x & y \\ 0 & 1 & z \\ 0 & 0 & 1 \end{bmatrix}\right) = \psi(x+z)$, and W_φ^θ is the (θ or) ψ-Fourier coefficient of φ along the maximal unipotent subgroup N of $GL(3)$.

In order to exploit the analytic properties of $\mathcal{Z}(s,\varphi)$ (and relate them to those of $L(s,\pi) = \Pi L(s,\pi_v)$) it is necessary to show that the L and ε factors which arise from the local versions of $\mathcal{Z}(s,\varphi)$ coincide with those defined in [GoJa] by way of the integrals $\mathcal{Z}(s,\Phi,f)$. This is a delicate enough technical problem which is further complicated by the fact that not all local representations have Whittaker models. Once settled, the converse theorem is easily established. Some uses of the converse theorem are reviewed in [Ge1]; although the "trace formula" ostensibly renders the converse theorem redundant in establishing "Langlands functionality", the theorem remains a satisfying tool for recognizing automorphic cuspidal representations (cf. Sections (1.6) and (1.7) below).

1.5. The metaplectic group.

Weil's metaplectic group enters naturally into the theory of L-functions for at least two reasons. Firstly, it provides the natural group theoretic home for the theta-functions which have played a continuous role in the theory since the time of Riemann. Secondly, Weil's realization of the metaplectic group as a group of unitary operators ("Weil's representation") makes it possible to bring to bear the full force of the theory of group representations on the construction of L-functions. In particular, Weil's realization of the metaplectic group $Mp_{\mathbb{A}}$ (a central extension of $Sp_n(\mathbb{A})$) provides the basis for Howe's theory of dual reductive pairs, and it is the latter theory which has so strongly influenced much of the recent analysis of L-functions by way of explicit integral formulas. We shall explain these matters in some detail in Chapter III(1.1) below.

For a general discussion of the metaplectic group and automorphic representations of dual reductive pairs, the reader is referred to [Ge3,4], [Lion-Vergne], [Howe], [Rallis 1-3], [MVW] and the references therein.

In [GePS2], Steps (1)-(5) of the local-global L-function machine described in Section I.1.7 are applied to the global Shimura type zeta-integral

$$\text{(II.1.5.1)} \qquad \int_{Z_{\mathbb{A}}GL_2(F)\backslash GL_2(\mathbb{A})} \varphi(g)\theta_\chi(g)E(g,s,f_s)dg = \mathcal{Z}(s,\varphi,\theta_\chi,\Phi) \; .$$

Here φ is a cusp form (of half-integral weight) belonging to the space of an automorphic cuspidal representation $\bar{\pi}$ of the metaplectic cover of $GL_2(\mathbb{A})$, θ_χ is essentially the classical theta-function θ_Ψ lifted to this metaplectic group, and $E(g,s,f_s) = \sum f_s^\Phi(\gamma g)$ is a (by now) familiar Eisenstein series on $GL_2(\mathbb{A})$. We note that the functions φ and θ_χ are indeed of *half-integral* weight, i.e., they do not factor through functions on $GL_2(\mathbb{A})$ itself. Nevertheless, the integral in (II.1.5.1) makes sense because the *product* of two such "genuine" metaplectic functions defines an ordinary automorphic form on $GL_2(\mathbb{A})$. (In classical terms, the product of two forms of half-integral weight is of integral weight.) The result is that we may attach to $\bar{\pi} = \otimes\bar{\pi}_p$ a "nice" L-function $L(s,\bar{\pi}) = \Pi L(s,\bar{\pi}_p)$ which by the converse theorem for $GL(2)$ is recognizable as the Jacquet-Langlands L-function of an automorphic cuspidal representation $\pi = \otimes\pi_p$ of GL_2. This is the generalized Shimura correspondence $\bar{\pi} \longrightarrow \pi$ developed in [GePS2].

A similar treatment of the Hecke type zeta-integrals

$$\mathcal{Z}(s,\varphi) = \int_{F^x\backslash\mathbb{A}^x} \varphi\left(\begin{bmatrix} a & 0 \\ 0 & 1 \end{bmatrix}\right) |a|^{s-1/2} d^x a \; , \qquad \varphi \in H_{\tilde{\pi}} \; ,$$

is given in [GePS3], but here the required Euler product expansion for $\mathcal{Z}(s,\varphi)$ is possible only for the so-called "distinguished" (or "exceptional") cusp forms $\bar{\pi}$. These $\bar{\pi}$ are characterized in [GePS3], and their L-functions are shown to reduce to Hecke's abelian L-series $L(s,\chi)$.

Finally, we must mention the important recent work of Waldspurger which is related not only to the Shimura correspondence above, but also to the more general theory of L-functions discussed in Chapter III. [Wald] describes Shimura's correspondence as a correspondence between automorphic representations of the ("dual") reductive groups $\overline{SL}_2(\mathbb{A})$ and PGL_2 (where $\overline{SL}_2(\mathbb{A})$ denotes the rank 1 metaplectic group realized as a *two* fold cover of $SL_2(\mathbb{A})$). Waldspurger's correspondence between φ in $H_{\tilde{\pi}}$ and F in H_π is given by way of the formula

$$F_\varphi(g) = \int_{SL_2(F)\backslash\overline{SL}_2(\mathbb{A})} \varphi(s)\theta(s,g)ds \tag{II.1.5.2}$$

where $\theta(s,g)$ is (the restriction to $\overline{SL}_2 \times PGL_2$ of) a theta-function on some bigger metaplectic group containing both PGL_2 as $\overline{SL}_2$ as mutually centralizing subgroups. This is precisely the kind of formula which is part of Howe's general theory; cf. Section III.1.1 below for more details. In the classical language of forms of half-integral or integral weight, formulas of this type first appeared in [Shin2] and [Niwa]. In [GePS4], it is shown directly that even the local correspondences of Waldspurger and Shimura coincide. More precisely, the "Mellin-transforms"

$$\mathcal{Z}(s,F_\varphi) \quad \text{and} \quad \mathcal{Z}(s,\varphi,\theta_\chi,\Phi)$$

coincide as Euler products.

1.6. Symmetric powers of L-functions.

Let us return now to Shimura's *symmetric square* L-function

$$L^2(s,f) = \Pi \frac{1}{(1-\alpha_p^2 p^{-s})(1-\alpha_p\beta_p p^{-s})(1-\beta_p^2 p^{-s})}$$

discussed in Section (II.1.3). Recall that Shimura's proof of the holomorphy of $L^2(s,f)$ proceeded by way of basic identity (II.1.3.4) relating $L^2(s,f)$ to $E(z,s)$. The natural generalization of the right-hand side of (II.1.3.4) is the global zeta-integral

$$\mathcal{Z}(s,\varphi,F,\Phi) = \int_{SL_2(F)\backslash SL_2(\mathbb{A})} \varphi(g)\theta_\Phi(g)E(g,s)dg$$

where φ belongs to the space H_π of an automorphic cuspidal representation π of $GL(2)$, $\theta(g)$ is a basic theta function lifted to the metaplectic group, and $E(g, s, F)$ is an Eisenstein series (of "half-integral weight") on the metaplectic group. Once again, the product of $E(g, s, F)$ with $\theta_\Phi(g)$ defines a slowly increasing automorphic form on $SL_2(\mathbb{A})$.

By applying (a suitable modification of) the "L-function machine" of Section I.(1.7) to this zeta integral, it was shown in [Ge Ja] that the resulting L-function

$$L^2(s, \pi) = \Pi L^2(s, \pi_p)$$

is (almost always) entire and satisfies a simple functional equation. If π happens to correspond to the classical cusp form f discussed above, then $L^2(s, \pi)$ =(a translate of) $L^2(s, f)$; more precisely, and more generally, if π_p is unramified and corresponds to the conjugacy class $t_p = \begin{bmatrix} \alpha_p & 0 \\ \beta_p & 0 \end{bmatrix}$ in $GL_2(\mathbb{C})$, then

$$L^2(s, \pi_p) = \frac{1}{det(I - r(t_p)p^{-s})} = L(s, \pi_p, r)$$

where $r : {}^L G \longrightarrow GL(3, \mathbb{C})$ is the *Adjoint-square* representation obtained from the adjoint action of $GL_2(\mathbb{C})$ on the Lie algebra of $SL_2(\mathbb{C})$. This explains the terminology "symmetric square". Indeed, when restricted to the torus, it is clear that the "adjoint square" differs from the "symmetric square" representation

$$Sym^2(\rho_2)\begin{bmatrix} \alpha_p & 0 \\ 0 & \beta_p \end{bmatrix} \longrightarrow \begin{bmatrix} \alpha_p^2 & 0 & 0 \\ 0 & \alpha_\beta\beta_p & 0 \\ 0 & 0 & \beta_p^2 \end{bmatrix}$$

only by the first power of the determinant. (Here ρ_2, as usual, denotes the standard representation of $GL_2(\mathbb{C})$.

Using the converse theorem *for* $GL(3)$, it is also shown in [Ge Ja] that $L^2(s, \pi)$ must be the L-function of an automorphic representation Π of $GL(3)$; this is the "Gelbart-Jacquet lifting from $GL(2)$ to $GL(3)$" which has found many applications in the theory of automorphic forms; cf. [La4] and the remarks below.

Clearly it would be desirable to similarly analyze the higher degree symmetric power L-functions $L(s, \pi, Sym^m(\rho_2)$ whose local components are (almost everywhere) equal to

$$\prod_{0 \leq j \leq m} \frac{1}{(1 - \alpha_p^j \beta_p^{m-j} p^{-s})}$$

if $t_{\pi_p} = \begin{bmatrix} \alpha_p & 0 \\ 0 & \beta_p \end{bmatrix}$. Even more desirable would be a proof of the existence of a lifting

$$Sym^m : \pi \longrightarrow \Pi$$

between automorphic cuspidal representations of $GL(2)$ and $GL(m+1)$ such that

$$\text{(II.1.6.1)} \qquad L(s, \pi, Sym^m(\rho_2)) = L(s, \Pi, \rho_{m+1}) \,.$$

See Section 1.8 below for a general discussion of such "liftings".

As discussed elsewhere, the existence of such a lifting would imply (among other things) the generalized Ramanujan-Petersson conjecture for π; cf. the concluding remarks of [Katz]. Unfortunately, very little is known (at the time of this writing) about the poles of $L(s, \pi, Sym^m)$, even when $m = 3, 4, 5$. What little is known results from the theory of [Sha9] discussed in Sections II.2 and III.2 below (*and* the Adjoint square lifting from $GL(2)$ to $GL(3)$).

1.7. $GL(n) \times GL(m)$.

In a long series of papers, Jacquet, Piatetski-Shapiro, and Shalika have been developing a theory of "Rankin-Selberg convolutions" for pairs of automorphic representations of $GL(n)$ and $GL(m)$. For $n = m = 2$, the complete theory appears in [Jacquet]; as already explained in Section (1.2) above, this theory is based on the original works of [Rankin] and [Selberg]. In general, the goal is to understand the analytic properties of the Euler products

$$L(s, \pi \times \pi') = \Pi L(s, \pi_p \times \pi'_p)$$

where for all unramified p,

$$L(s,\pi_p \times \pi'_p) = det(1-(t_p \otimes t'_p)p^{-s})^{-1} \ .$$

Although complete results on the meromorphic continuation and functional equation of these L-functions have not yet been published, many interesting preliminary results about the zeroes and poles of $L(s,\pi \times \pi')$ have already found important applications in the literature.

For example, by analyzing certain zeta-integrals for $L_S(s,\pi\times\pi')$ (similar to the local and global integrals $I(s,\varphi_1,\varphi_2,E)$ used in [Jacquet] for $GL(2)\times GL(2)$, it is shown in [JS1] that two irreducible unitary cuspidal representations π and π' of $GL_n(\mathbb{A})$ are equivalent of and only if

$$L(s,\pi'_p \times \tilde{\pi}_p) = L(s,\pi'_p \times \tilde{\pi}'_p)$$

for each (finite) p such that π_p and π'_p are unramified. This result immediately implies a "strong multiplicity one" theorem for GL_n: if two unitary cuspidal representations π and π' are equivalent almost everywhere, then $\pi \approx \pi'$. For refinements of this strong multiplicity result, as well as a discussion of the problem's history, see [Moreno1].

A still more remarkable result is the generalization of this multiplicity one theorem to any irreducible automorphic (not necessarily cuspidal) representation of $GL_r(\mathbb{A})$. More precisely, let $P = MU$ be a "standard" parabolic subgroup of $G = GL_r$. Assume $M = GL_{r_1} \times \ldots \times GL_{r_u}$. Let $\sigma = \sigma_1 \otimes \ldots \otimes \sigma_u$ be a cuspidal automorphic representation of $M_{\mathbb{A}}$ and consider

$$\xi = Ind(G_{\mathbb{A}},P_{\mathbb{A}},\sigma) = \otimes\xi_v \ .$$

Every irreducible automorphic representation of $G_{\mathbb{A}}$ is a constituent of such a ξ for some P and σ; cf. [La8]. Now let Q, of type $(s_1,\ldots,s_w)$, be another standard parabolic subgroup of G. Fix a cuspidal representation $\tau = \tau_1\otimes\ldots\otimes\tau_w$ of $GL_{s_1}(\mathbb{A})\times\ldots\times GL_{s_w}(\mathbb{A})$ and let $\eta = Ind(G_{\mathbb{A}},\mathbb{Q}_{\mathbb{A}},\tau)$. We shall say (σ,P) and (τ,Q) are associate if and only if 1) $u = w$, 2) there exists a permutation ϕ in u letters such that $s_j = r_{\phi(j)}$, and

3) $\tau_j \cong \sigma_{\phi(j)}$, $1 \leq j \leq u$. Then the following "classification theorem" (cf. [La8]) is proved in [JS2] (Theorem 4.4):

__Theorem.__ Let P, Q, $\sigma_j = \otimes_v \sigma_{jv}$, and $\tau_k = \otimes_v \tau_{kv}$ be as above. Fix a finite set of places S, containing the archimedean ones such that for $v \notin S$, σ_{jv} and τ_{kv} are both unramified, $1 \leq j \leq u$, $1 \leq k \leq w$. Then ξ_v and η_v have the same unramified component for $v \notin S$ if and only if (σ, P) and (τ, Q) are associate.

Note that the case $P = Q = G$ reduces to the multiplicity one result just discussed. The proof is a beautiful application of the following analytic properties of the L-functions $L_S(s, \pi \times \pi')$ proved in [JS 1,2].

a) Assume $n \neq m$; then $L_S(s, \pi \times \pi')$ has a continuous extension to the closed half plane $Re(s) \geq 1$ (Proposition 3.3 of [JS2]).

b) Assume $n = m$. Let X be the set of s on $Re(s) = 1$ such that $\alpha^{s-1} \otimes \pi$ is equivalent to the contragredient of π'. Then $L_S(s, \pi \times \pi')$ has a pole at s_0 on $Re(s) = 1$ if and only if $s_0 \in X$. This pole is then simple (Proposition 3.6 of [JS2]). Moreover, $L_S(s, \pi \times \pi')$ extends to a continuous function of s in the complement of X in the half space $Re(s) \geq 1$.

c) For $Re(s) > 1$, the Euler product $L_S(s, \pi \times \pi')$ converges absolutely (Theorem 5.3 of [JS1]). This, together with Theorem 5.1 of [Sha3] (cf. [JS3] if $m = 1$), implies that $L_S(s, \pi \times \pi')$ is in fact non-zero for $Re(s) \geq 1$ (Theorem 3.7 of [JS2]); cf. Section II.2.6 below for more on the non-vanishing of automorphic L-functions.

Proofs of most of these facts are again based on the analytic properties of the integral representation $I(s, \varphi_1, \varphi_2, E)$ for $L_S(s, \pi \times \pi')$, and the known analytic properties of E.

Finally, we mention that a complete *local* theory for $L(s, \pi \times \pi')$ is developed (for non-archimedean places) in [JPSS2], i.e., steps (3)-(5) of the L-function machine are carried out. The local *and* global theory is treated from a completely different point of view in [Sha3, 4, 7, 8], which we shall soon (finally) begin to discuss in earnest.

1.8. Additional notes and references: L-functions and the lifting problem.

Langlands' functoriality conjecture, also known as the "lifting problem for automorphic forms", may be explained as follows (cf. [La3] and [Bo1]).

If G and H are two connected reductive algebraic groups over a number field F, denote by $\mathcal{A}(G)$ and $\mathcal{A}(H)$ the corresponding sets of irreducible automorphic representations of $G(\mathbb{A}_F)$ and $H(\mathbb{A}_F)$ respectively. The functoriality conjecture asserts that to each homomorphism $\theta : {}^LG \longrightarrow {}^L H$ there exists a "lifting" $\theta^* : \mathcal{A}(G) \longrightarrow \mathcal{A}(H)$ satisfying the following properties: if $\pi = \otimes_v \pi_v$ and $\theta^*(\pi) = \otimes_v \theta^*(\pi_v)$, then for any finite dimensional complex representation r of ${}^LH$, and any non-trivial additive character $\psi = \Pi \psi_v$ of $F \backslash \mathbb{A}$,

$$L(s, \pi_v, r \cdot \theta \cdot \eta_v^G) = L(s, \theta^*(\pi_v), r \cdot \eta_v^H) \tag{II.1.8.1}$$

and

$$\varepsilon(s, \pi_v, r \cdot \theta \cdot \eta_v^G, \psi_v) = \varepsilon(s, \theta^*(\pi_v), r \cdot \eta_v^H, \psi_v) .$$

Here $\eta_v^G : {}^LG_v \longrightarrow {}^L G$ and $\eta_v^H : {}^LH_v \longrightarrow {}^L H$ are the natural homomorphisms (and the original L-group morphism θ is assumed to be nice with respect to the Galois group Γ_F).

Let us note first that this Conjecture implies (in principle) the main Conjecture L of this paper. Indeed, since any finite-dimensional representation $r : {}^LG \longrightarrow GL_d(\mathbb{C})$ may also be regarded as an L-group morphism from ${}^LG$ to the L-group ${}^LH$ of $H = GL_d$, the functoriality conjecture reduces the problem of establishing the analytic properties of the automorphic L-function $L(s, \pi, r)$ on G to that of the (known) standard L-function $L(s, \theta^*(\pi), \rho_d)$ on GL_d.

Let us also give a brief list of examples where the functoriality conjecture is known, making no pretense of completeness.

(a) Let $G = GL_2$, $H = GL_3$, and let $\theta : GL_2(\mathbb{C}) \longrightarrow GL_3(\mathbb{C})$ denote the map obtained by composing projection onto $PGL_2(\mathbb{C})$ with the natural adjoint representation $Ad: PGL_2(\mathbb{C}) \longrightarrow GL_3(\mathbb{C})$. Then, as already remarked in §II.1.6, the existence of θ^* was established in [Ge Ja]; subsequently, the existence of θ^* was discussed in [Flicker2] using the trace formula.

(b) Let $G = U_{2,1}$ and $H = Res_{E/F}GL_3$, where E/F is the quadratic extension of number fields by which the unitary group $U_{2,1}$ is defined. If

$$\theta : GL_3(\mathbb{C}) \rtimes Gal(E/F) \longrightarrow (GL_3(\mathbb{C}) \times GL_3(\mathbb{C})) \rtimes Gal(E/F)$$

is defined by $\theta(g \rtimes \tau) = (g, g) \rtimes \tau$, then it is the subject matter of [Rogawski] to show that θ^* exists; cf. [GPS1].

(c) "Stablization" of the trace formula, as suggested first in [Labesse La], is expected to lead to a proof of the functoriality conjecture in a large number of cases; cf. [La10].

(d) Finally, we concentrate on the important example of "base change" for GL_n. Let E be a cyclic extension of prime degree ℓ of a number field F. We let $G = GL_n$ and $H = Res_{E/F}G$, both regarded as groups over F. Then ${}^LG = GL_n(\mathbb{C}) \times Gal(E/F)$, whereas ${}^LH = \underbrace{GL_n(\mathbb{C}) \times \ldots \times GL_n(\mathbb{C})}_{\ell}$ $\rtimes Gal(E/F)$, and the generator τ of $Gal(E/F)$ acts by the permutation $(1\ 2 \ldots \ell)$. Now define $\theta : {}^LG \longrightarrow {}^LH$ by $\theta(g \times \tau) = (g, g, \ldots, g) \rtimes \tau$. If θ^* exists, then for π in $\mathcal{A}(G)$, $\theta^*(\pi)$ in $\mathcal{A}(H)$ is called its *base change lift* to $GL_n(\mathbb{A}_E)$.

For $n = 2$, the existence of θ^* is due to Saito, Shintani, and Langlands (cf. [La9]), and the case $n = 3$ was later treated in [Flicker3]. The existence of θ^* for general n is the subject matter of a recent preprint of Arthur and Clozel [Art Cℓ]. We note that if Π and T are base change lifts of π in $\mathcal{A}(GL_n(\mathbb{A}_F))$, and $\tau \in \mathcal{A}(GL_m(\mathbb{A}_F))$, then

$$L_S(s, \Pi \times T) = \coprod_{i=1}^{\ell} L_S(s, \pi \times (\tau \otimes \eta^i)),$$

where the L-functions on both sides are those of [JPSS2] for $GL_n \times GL_m$ and η is a character of $F^x N_{E/F}(\mathbb{A}_E^x) \backslash \mathbb{A}_F^x$.

A local base change lifting, which is compatible with the global one discussed, also exists, as do identities, including one for the "root numbers" ε, similar to the ones above.

In closing, we mention that L-functions closely related to those defined for $GL(n) \times GL(n)$ are obtained from the following "restriction of scalars" data. Fix $G = Res_{E/F} GL_n$ with $\ell = [E:F] = 2$ so that ${}^L G = (GL_n(\mathbb{C}) \times GL_n(\mathbb{C})) \times Gal(E/F)$; let π be an arbitrary automorphic representation of $G_{\mathbb{A}}$ (i.e. not necessarily a base change lift), and define

$$r : {}^L G \longrightarrow GL_{2n}(\mathbb{C})$$

to be the "twisted tensor" representation given by $r(g_1, g_2) = g_1 \otimes g_2$ and $r(\tau)(x \otimes y) = y \otimes x$. For the case $n = 2$ and $F = \mathbb{Q}$, the analytic properties of the corresponding L-function were obtained by [Asai]; for arbitrary n and F, see [Flicker1]. Analogous results for $\ell = [E:F] = 3$ are especially intriguing; these are the subject matter of [PSR3] discussed in III.1.4 below. We note that in case $n = 2$, the L-functions of Asai and Jacquet (for $GL(2) \times GL(2)$) are distinct, but might agree at the local places v which split in E.

1.9. Concluding remarks.

In Sections 1 through 8 of this Chapter we have discussed several different methods of using integral expressions ("zeta-integrals") to define and analyze automorphic L-functions of the form

$$L(s, \pi, r) = \Pi L(s, \pi_v, r_v) .$$

We have referred to these methods as "Tate's method", "Hecke's method", the "method of Rankin-Selberg", and "Shimura's method". We have seen how several authors developed these methods in treating $L(s, \pi, r)$ for $G = GL(2)$, $GL(2) \times GL(2)$, $GL(n)$ and $GL(n) \times GL(m)$, with r the standard representation of ${}^L G$, or its symmetric square.

It should be mentioned that a general approach to this problem, which anticipated if not actually incorporated several of these methods, was already suggested fifteen years ago by Piatetski-Shapiro in [PS2]. This is a paper full of ideas which are only now beginning to be fully understood (cf. the work of Piatetski-Shapiro and collaborators discussed in Chapter III.1 below). A more recent survey of these ideas, taking into account the subsequent works of Shimura, appears in [PS3] and [Bump2].

We close with some remarks about the different ways Eisenstein series and Whittaker models are used in these various zeta-integral methods.

(1) So far, Eisenstein series enter the picture here only in the Rankin-Selberg or Shimura type zeta-integrals; soon we shall see that Piatetski-Shapiro and Rallis' generalization of the Godement-Jacquet integrals (Tate's method) also includes Eisenstein series in the global integrand. However, these uses of Eisenstein series are very different from the way L-functions and Eisenstein series are related in Langlands' original approach ([La1]). There the L-functions describe the constant term of some Eisenstein series defined on a larger group (or even the first Fourier coefficient of these Eisenstein series in the more recent work of Shahidi discussed below).

(2) As far as Whittaker models are concerned, we note that π is assumed to have such a model in all the methods discussed so far (except for that of Tate-Godement-Jacquet). This will also be the case for the refinement of Langlands' approach discussed in Section II.2 below.

(3) Finally, we mention some interesting *local* zeta-integrals analyzed in [Rodier]. These integrals essentially generalize the local Mellin transforms

$$\int_{F_v^x} W\begin{pmatrix} a & & 0 \\ & 1 & \\ & & \ddots \\ 0 & & 1 \end{pmatrix} |a|^{s'} d^x a$$

from $GL(n)$ to a fairly general *split* reductive group G, and fairly general representations of its L-group, *provided* π_v is the unique non-degenerate constituent of a principal series representation $Ind_B^G \chi$. The only problem here is that we have no idea how these local factors are related to any reasonable *global* zeta-integrals. In other words, we have no relevant global framework for these computations, and hence no way to put into operation the L-function machine described in I.1.7.

§2. EISENSTEIN SERIES AND GENERIC REPRESENTATIONS

We return to the theory of Langlands sketched in Section 2 of Chapter I, in particular the crucial relation (I.2.5.3) which asserts that

$$\text{(I.2.5.3)} \quad M(s,\pi)f = \left(\bigotimes_{v\in S} A(s,\pi_v)f_v\right) \otimes \bigotimes_{v\notin S} \tilde{f}_v^0 \prod_{i=1}^{m} \frac{L_S(is,\pi,\tilde{r}_i)}{L_S(1+is,\pi,\tilde{r}_i)} \,.$$

Recall here that $\pi = \otimes\pi_v$ is an automorphic cuspidal representation of the reductive group G, $r_1,\ldots,r_m$ are certain basic representations of ${}^LG$, and $M(s,\pi)$ is an intertwining operator appearing in the constant term of an Eisenstein series defined on $H_{\mathbb{A}} \supset P = G_{\mathbb{A}}U^P$ (by induction from the representation $\pi \otimes exp\ (H_P(\cdot)(s))$). The point was that the known meromorphic continuation of the Eisenstein series (or $M(s,\pi)$) can imply a meromorphic continuation for the automorphic L-functions $L_S(s,\pi,\tilde{r}_i)$. However, as we already explained in the case when the Jacquet-Langlands L-function $L(s,\pi)$ occurs in (I.2.5.3), this relation offers no additional information on the location of poles (if any) for $L(s,\pi,r_i)$, or the best possible functional equation.

In [Sha1-3] one proceeds further by assuming that π is *non-degenerate* i.e., possesses a Whittaker model. In this case, one can compute non-trivial "Fourier coefficients" of the resulting Eisenstein series $E(s,f,h)$, and relate them directly to the L-functions $L(s,\pi,r_i)$ (without the appearance of quotients).

Since we have discussed Fourier expansions only briefly, and then in only the special case of $GL(2)$, we begin by recalling some basic definitions and facts from the general theory of Whittaker models.

2.1. Whittaker models: General notions.

Everything in this section makes sense for an arbitrary quasi-split connected reductive group over an $\mathbb{A}$-field F, but to ease the exposition, we shall assume that the reductive group H containing our G is split (and hence dispense with the need to bring into play the Galois group of F when discussing roots, ${}^LG$, etc.).

Fix a Borel subgroup $B = TU$ of H with T a maximal torus and U a maximal unipotent subgroup of H. An F-morphism f from U to $\bar{F}$ is called *non-degenerate* or *generic* if for $u = \Pi_{\alpha \in \Phi^+} x_\alpha(t_\alpha)$, we have

$$f(u) = \sum K_\alpha t_\alpha \tag{II.2.1.1}$$

with all K_α in F *non-zero*. Here we are using x_α to denote the root subgroup map from $\bar{F}$ to U corresponding to the positive root α in Φ^+. Thus, for $H = GL(n) \supset U = \left\{ \begin{bmatrix} 1 & & x_{ij} \\ & \ddots & \\ 0 & & 1 \end{bmatrix} \right\}$, condition (II.2.1.1) means that f is non-trivial on each of the simple root subgroups $\left\{ \begin{bmatrix} 1 & & 0 & & 0 \\ & \ddots & & & \\ & & 1 & x_{i,i+1} & \\ & & & \ddots & 0 \\ 0 & & 0 & & 1 \end{bmatrix} \right\}$ In general, we may extend f to a map of $U(F_v)$ or $U(\mathbb{A}_F)$.

Given a character ψ_U of $U(F_v)$ (or $U(F)\backslash U(\mathbb{A}_F)$), we call ψ_U *non-degenerate* or *generic* if $\psi_U(u) = \psi(f(u))$ with $f : U \longrightarrow \bar{F}$ non-degenerate and ψ a fixed non-trivial (additive unitary) character of F_v (or $F\backslash\mathbb{A}_F$). Henceforth, we shall often drop the subscript U in describing the character ψ_U of U, and trust the reader will be able to distinguish between such a character and a character ψ_F of F. In the local case we call ψ_v *unramified* on $U(F_v)$ if every $K_\alpha \in O^*_{F_v}$ and the largest ideal on which ψ_F is trivial is O_{F_v}. If $\psi = \Pi_v \psi_v$ is a generic character of $U(A)\backslash U(\mathbb{A})$, then for each v, ψ_v is a generic character of U_v, and almost all of the ψ_v's are unramified.

Given an automorphic cuspidal representation $\pi = \otimes \pi_v$ of $G_{\mathbb{A}}$, and a function φ in the space H_π of π, consider the ψ-th Fourier coefficient

$$W^0_\varphi(g) = \int_{U^0(F)\backslash U^0(\mathbb{A})} \varphi(ug)\overline{\psi^0(u)}du$$

where $U^0 = U \cap G$, and ψ^0 is the restriction of ψ_U to U^0. We say π is *ψ-generic* if $W^0_\varphi \not\equiv 0$ for some φ; the representation π is simply *generic* if it is ψ-generic for some generic ψ. In this case, each component π_v is ψ_v-generic, i.e., there exists a non-zero linear functional λ on the space H_{π_v} such that $\lambda(\pi(u)w) = \psi_v(u)\lambda(w)$ for all $u \in U^0_v$ and $w \in H_{\pi_v}$. (This follows immediately from the fact that the evaluation map $\varphi \longrightarrow W^0_\varphi(e)$ defines such a functional on H_π.) Such functionals are called ψ_v-*Whittaker functions.* Consequently, each π_v has a canonical *Whittaker model* $\mathcal{W}(\pi_v, \psi_v)$.

In more detail, let H^*_v denote the complex vector space of Whittaker functionals on H_{π_v}. In [Shalika] and [GelKaz] it is shown that π_v irreducible unitary (admissible) implies $dim\ H^*_v \leq 1$; so since π_v has just been observed to be generic, this implies $dim\ H^*_v = 1$. For each v in H_{π_v} define a *Whittaker function* W_v on G_v by

$$W_v(g) = \lambda(\pi_v(g)v) .$$

Then $W_v(ug) = \psi_v(u)W_v(g)$ for $u \in U^0_v$. Let $\mathcal{W}(\pi_v, \psi_v)$ denote the vector space of all Whittaker functions, on which G_v acts by right translation; this space $\mathcal{W}(\pi_v, \psi_v)$ is unique, and called the *Whittaker model* for π_v. (If $F_v = \mathbb{R}$ or $\mathbb{C}$, we consider only Whittaker functionals which are defined on the subspace of *smooth* vectors in H_{π_v} and *continuous* with respect to the appropriate topology; cf. Section 3 of [Shalika] for details.)

2.2. Whittaker models for $I(s, \pi_v)$.

Next we consider Whittaker models for the representations $I(s, \pi_v)$ on H_v induced from $\pi \otimes exp(H_P(\cdot)(s)) \otimes 1$ on $P_v = G_v U^{P_v}$. For the case $H = SL(2)$, we have the familiar Whittaker functional defined on the space of $Ind\ \chi_p |a|^s_p$ by

(II.2.2.1)

$$f \longrightarrow \lambda(s, \chi_p, \psi_p)(f) = \int_{F_p} f\left(\begin{pmatrix} 0 & 1 \\ -1 & 0 \end{pmatrix}\begin{pmatrix} 1 & x \\ 0 & 1 \end{pmatrix}\right)\overline{\psi_p(x)}dx .$$

The convergence of the integral defining this Whittaker functional is discussed in detail in [God], Section 1.9. In general, one can define a Whittaker functional $\lambda(s, \pi_v, \psi_v)$ on the space of $I(s, \pi_v)$ by means of a similar type of integral $\lambda(s, \pi_v, \psi_v)(f)$. For p-adic fields, it is known (cf. [Sha3] and [CS]) that this integral converges for all s (but not absolutely), and defines an entire function of s for each fixed f. Moreover, there exists an f for which $\lambda(s, \pi_v, \psi_v)f \neq 0$, i.e. λ is a well-defined *non-trivial* Whittaker functional . For *infinite* places, the analogue of this result is the subject matter of [Sha2]. If for each f in the space of $I(s, \pi_v)$ we define a function W_f on G_v by

$$W_f(g) = \lambda(s, \pi_v, \psi_v)(I(s, \pi_v)(g)f)$$

then the space

$$\mathcal{W}(s, \pi_v, \psi_v) = \{W_f : \ f \text{ in } I(s, \pi_v)\}$$

comprises the ψ_v-Whittaker model for $I(s, \pi_v)$.

2.3. Fourier coefficients of Eisenstein series.

We compute finally the ψ-th Fourier coefficient of the Eisenstein series

$E(s, P, f, h)$ introduced in I.(2.5.2). Recall that

$$E(s, P, f, h) = \sum_{\gamma \in P_F \backslash H_F} f_s(\gamma h)$$

where $f_s \in Ind(s, \pi)$. Assuming still that π is ψ-generic, let

$$E_\psi(s, f, P, h) = \int_{U(F)\backslash U(\mathbb{A})} E(s, f, P, uh)\overline{\psi_U(u)}du$$

denote the corresponding ψ-th Fourier coefficient of $E(s, f, P, h)$ (in contrast to a *constant* term $\int E(s, f, P, uh)du$). Choose a finite set of places S of F, including all the archimedean ones, such that for $v \notin S$, π_v, $H$and ψ_v are all unramified. Assume also that f is of the form $\Pi_v f_v$, where each f_v is in the space of $I(s, \pi_v)$, and for all $v \notin S$, f_v is the unique K_v-fixed function such that $f_v(e_v) = 1$. Then by computations in Section 5 of [Sha3],

$$\text{(II.2.3.1)} \qquad E_\psi(s, f, P, e) = \prod_{v \in S} W_v(e_v) \prod_{i=1}^{m} L_S(1 + is, \pi, \tilde{r}_i)^{-1}$$

where W_v stands for W_{f_v}.

This formula is much more useful than its predecessor I.(2.5.3) which relates $M(s,\pi)f$ to the *quotient* of L-functions $L_S(s,\pi,r_i)$. First of all, (2.3.1) might be used right away to read off information about the zeros of $L(s,\pi,r_i)$ in terms of the poles of $E(s)$. Secondly, the functional equation for $E(s,f,P,h)$ has now a better chance of yielding the sought-after equation for $L(s,\pi,r_i)$. To obtain this equation we need the so-called "local coefficients" introduced in [Sha3].

2.4. Local coefficients and the functional equation for $L_S(s,\pi,r)$.

For each place v of F, let $A(s,\pi_v)$ denote the intertwining operator associated to $I(s,\pi_v)$ so that $\otimes A(s,\pi_v) = M(s,\pi)$. Although initially defined by an integral convergent only in some half-plane, $A(s,\pi_v)f_v$ is known to define a meromorphic function of s in $\mathbb{C}$ (cf. [Harish-Chandra] and [Silberger]).In fact, when v is non-archimedean, $A(s,\pi_v)f_v$ is a rational function in q_v^{-s}. In general, it makes sense to define a Whittaker functional $\lambda'(s,\psi_v,\pi_v)$ on the image of $I(s,\pi_v)$ by $A(s,\pi_v)$ by using a formula similar to the one defining $\lambda(s,\pi_v,\psi_v)$ (cf. [Sha3]). Thus it follows from the uniqueness of Whittaker functionals that

$$\text{(II.2.4.1)} \qquad \lambda(s,\pi_v,\psi_v)(f_v) = C_{\psi_v}(s,\pi_v)\lambda'(s,\psi_v,\pi_v)(A(s,\pi_v)(f_v))$$

for some complex number $C_{\psi_v}(s,\pi_v)$. These are the *local coefficients* of [Sha3]. Because of the properties of Whittaker functionals and intertwining operators just recalled, it follows from this definition that the $C_{\psi_v}(s,\pi_v)$ are meromorphic functions of s, in fact rational in q_v^{-s}, for v non-archimedean. It can also be checked that

$$\text{(II.2.4.2)} \qquad C_{\bar{\psi}_v}(\bar{s},\tilde{\pi}_v) = \overline{C_{\psi_v}(s,\pi_v)}\ .$$

One of the main results of [Sha3] is the functional equation

$$\text{(II.2.4.3)} \qquad \prod_{i=1}^{m} L_S(is,\pi,r_i) = \prod_{v\in S} C_{\bar{\psi}_v}(s,\tilde{\pi}_v) \prod_{i=1}^{m} L_S(1-is,\pi,\tilde{r}_i)$$

where $\tilde{\pi}_v$ is the $\bar{\psi}_v$-generic contragredient of π_v. This is a consequence of (2.3.1) and the functional equation for $E(s, f, P, h)$. Note the similarity between this general functional equation and the equation (I.1.5.3) established for $GL(2)$ using zeta-integrals.

Example 2.4.4. (*Derivation of the formulas (II.2.3.1) and (II.2.4.3) in case $H = SL(2)$).*

We simply compute from the definitions:

$$E_\psi(s, f, e) = \int_{N_{\mathbb{Q}}\backslash N_{\mathbb{A}}} E(s, f, n)\overline{\psi(n)}dn$$

$$= \int_{N_{\mathbb{Q}}\backslash N_{\mathbb{A}}} \left(\sum_{\gamma \in B_{\mathbb{Q}}\backslash H_{\mathbb{Q}}} f_s(\gamma n) \right) \overline{\psi(n)}dn$$

$$= \int_{N_{\mathbb{Q}}\backslash N_{\mathbb{A}}} f_s(n)\overline{\psi(n)}dn + \int_{N_{\mathbb{Q}}\backslash N_{\mathbb{A}}} \left(\sum_{\gamma \in N_{\mathbb{Q}}} f_s(w\gamma n) \right) \overline{\psi(n)}dn$$

(via Bruhat's decomposition)

$$= f_s(e) \int_{N_{\mathbb{Q}}\backslash N_{\mathbb{A}}} \overline{\psi(n)}dn + \int_{N_{\mathbb{A}}} f_s\left(\begin{pmatrix} 0 & 1 \\ -1 & 0 \end{pmatrix} n \right) \overline{\psi(n)}dn \ .$$

Because $\psi(n)$ is a non-trivial character of $N_{\mathbb{Q}}\backslash N_{\mathbb{A}}$, the first term equals zero, and we have

$$E_\psi(s, f, e) = \prod_p \int_{\mathbb{Q}_p} f_p \begin{pmatrix} 0 & 1 \\ -1 & 0 \end{pmatrix} \begin{pmatrix} 1 & x \\ 0 & 1 \end{pmatrix} \overline{\psi(x)}dx \ ,$$

with almost every f_p the unique normalized K_p invariant function f_p^0 in $I(s, \chi)$. In particular, if S is as in (II.2.3.1), then according to the definition (II.2.2.1), we have

$$E_\psi(s, f, e) = \left(\prod_{p \in S} W_{f_p}(e) \right) \left(\prod_{p \notin S} \int_{\mathbb{Q}_p} f_p^0 \left(w \begin{pmatrix} 1 & x \\ 0 & 1 \end{pmatrix} \right) \overline{\psi_p(x)}dx \right)$$

and it remains only to compute this last integral. For this, we recall that ψ_p "unramified" outside p in S means that $\mathbb{Z}_p$ is the largest ideal of $\mathbb{Q}_p$

on which ψ_p is trivial, and therefore

$$\int_{|x|=p^n} \overline{\psi_p(x)}dx$$

$$= \int_{p^{-n}\mathbb{Z}_p} \overline{\psi_p(x)}dx - \int_{p^{-(n-1)}\mathbb{Z}_p} \overline{\psi_p(x)}dx = \begin{cases} 0 & \text{if} \quad n > 1 \\ -1 & \text{if} \quad n = 1 \,. \end{cases}$$

It follows that

$$\int_{\mathbb{Q}_p} f_p^0\left(w\begin{pmatrix} 1 & x \\ 0 & 1 \end{pmatrix}\right)\overline{\psi_p(x)}dx$$

$$= \int_{|x|\leq 1} f_p^0\left(w\begin{pmatrix} 1 & x \\ 0 & 1 \end{pmatrix}\right)dx + \int_{|x|>1} f_p^0\left(n_x\begin{pmatrix} x^{-1} & 0 \\ 0 & x \end{pmatrix}k_x\right)\overline{\psi_p(x)}dx$$

$$= 1 + \sum_{n=1}^{\infty} \chi(p)|p^n|^{-(s+1)} \int_{|x|=p^n} \overline{\psi_p(x)}dx = 1 - \chi(p)p^{-(s+1)}$$

and thus that

$$E_\psi(s, f, e) = \left(\prod_{p\in S} W_{f_p}(e)\right) L_S(1 + s, \chi)^{-1} \,, \qquad \text{(II.2.4.5)}$$

as required by formula (II.2.3.1).

Remark. This computation consisted of two steps: first the Bruhat decomposition was used to describe $E_\psi(s, f, e)$ as $\Pi W_{f_v}(e)$and then $W_{f_v^0}$ was computed explicitly. In general, i.e., for arbitrary reductive H, one can proceed the same way, since only the "longest" Weyl group element will contribute non-trivially, and then the computation of $W_{f_v^0}$ is precisely the subject matter of [CS].

As regards the functional equation, we start with the functional equation for $E(s, f, e)$, namely

$$E(s, f, e) = E(-s, M(s)f, e) \,.$$

Then we compute

$$E_\psi(-s, M(s)f, e)$$

$$= \left(\prod_{p\in S} W_{A(s)f_p}(e)\right) \times \prod_{p\notin S} \int_{\mathbb{Q}_p} A(s, \chi_p) f_p^0\left(w\begin{pmatrix} 1 & x \\ 0 & 1 \end{pmatrix}\right)\overline{\psi_p(x)}dx$$

where $A(s,\chi_p)f_p^0 = \frac{L(s,\chi_p)}{L(s+1,\chi_p)}\tilde{f}_p^0$, i.e.,

$$E_\psi(-s, M(s)f, e) = \left(\prod W_{A(s,\chi_p)f_p}(e)\right) \frac{L_S(s,\chi)}{L_S(s+1,\chi)} \cdot \frac{1}{L_S(1-s,\chi^{-1})} \, .$$

The functional equation for $E(s,f,e)$ implies therefore (together with (II.2.4.5)) that

$$L_S(s,\chi) = \left(\prod_{p\in S} \frac{W_{f_p}(e)}{W_{A(s)f_p}(e)}\right) L_S(1-s,\chi^{-1}) \, .$$

According to the definition of the local coefficients $c_{\psi_v}(s,\pi_v)$ (cf. (II.2.4.1) above), we have

$$W_{f_p}(e) = c_{\psi_p}(s,\chi_p) W_{A(s)f_p}(e) \, ,$$

hence the sought-after functional equation

$$L_S(s,\chi) = \left(\prod_{p\in S} c_{\psi_p}(s,\chi_p)\right) L_S(1-s,\chi^{-1}) \, .$$

Remark. In addition to obtaining a (new) proof of the functional equation of Dirichlet's L-series (by way of the theory of Eisenstein series), we also obtain (by way of relation (II.2.4.5) and the known holomorphy of $E(s,h,e)$ on the imaginary axis) the famous result that $L_S(s,\chi)$ never vanishes on the line $Re(s)=1$; cf. (2.6) below for a vast generalization of this result.

2.5. Examples.

Here are some typical general applications of the functional equation (II.2.4.3).

First, consider example (xv) on page 50 of [La1], already described by Langlands as "particularly striking". The diagram data is:

$m=2:$	$a_1 = 10$	$\lambda_1 = 0$
	$a_2 = 5$	$\lambda_2 = 3\delta_1$

Thus H is of type G_2, G may be taken to be PGL_2, and our representations r_i of ${}^LG = SL_2(\mathbb{C})$ are r_1=the trivial representation and r_2=the symmetric cube representation (whose highest weight is $3\delta_1$).

For π any automorphic cuspidal representation of $PGL_2(\mathbb{A})$, the product in (II.2.4.3) reduces to

$$L_S(2s,1)L_S(s,\pi,Sym^3)$$

where the first L-function is just a partial Riemann zeta-function. (Recall that our m in (II.2.4.3) corresponds to Langlands' integer r, and the parameter s may be normalized so that each a_i equals i.) Thus the functional equation for $L_S(s,\pi,Sym^3)$ is an immediate consequence of (2.4.3) and the known functional equation for $L_S(s,1)$; this is the main result of [Sha1].

Next consider example (iv) on p. 48 of [La1]:

2	2		2	2 1
0—0		· · · · · ·	0—0═0	
α_1	α_2		α_{n-2}	α_{n-1}
$m=1$		$a_1 = 2n$		$\lambda_1 = 2\delta_1$

Here H is a simple split algebraic group of type B_n with Dynkin diagram 0—0—0 · · · · · · 0—0═0, whereas $G = GL(n)$ corresponds to the diagram of labelled (long) roots $\underset{\alpha_1}{0}$—0……0—$\underset{\alpha_{n-1}}{0}$; the maximal parabolic subgroup P of H has Levi component G, and the unipotent radical of ${}^LP \subset {}^LH = Sp_n(\mathbb{C})$ is isomorphic to the space of symmetric $n \times n$ matrices. In particular $r = r_1$ is $Sym^2(\rho_n)$, and the L-function which appears (alone) in (2.4.3) is $L_S(s,\Pi,Sym^2(\rho_n))$. Thus one immediately obtains a meromorphic continuation and functional equation for these important L-functions. (More precise results for these and related L-functions are discussed in §2.5 of Chapter III.)

Finally, we describe a less direct application of (II.2.4.3). For this, fix $n = 3$ in the example above, and assume Π is an automorphic cuspidal representation of $GL_3(\mathbb{A})$ which is the Gelbart-Jacquet lift of an automorphic cuspidal π on $PGL_2(\mathbb{A})$. Then (by Langlands' functoriality; cf. (II.1.8.1))

$$L(s,\pi,Sym^2(Sym^2(\rho_2)) = L(s,\Pi,Sym^2(\rho_3)) \ .$$

On the other hand, a classical computation (cf. Lemma 4.1.1 of [Sha3]) shows that

$$Sym^2(Sym^2(\rho_2)) = Sym^4(\rho_2) \oplus (\Lambda^2(\rho_2)) \ .$$

Therefore we must have

$$L(s,\Pi,Sym^2(\rho_3)) = L(s,\pi,Sym^4(\rho_2))L(s,1) \ ,$$

where $L(s,1)$ is the simplest possible Hecke L-function. Using Hecke's functional equation and the functional equation for $L_S(s,\Pi,Sym^2(\rho_3))$ (discussed in the last paragraph) we obtain a functional equation for $L(s,\pi,Sym^4(\rho_2))$. Similarly we can deal with the 5-th symmetric power of ρ_2. For details, see §4.1 of [Sha3]; both these examples are somewhat unexpected, since neither of the L-functions appears directly as an $L(s,\pi,r_i)$.

2.6. On the non-vanishing of automorphic L-functions for $Re(s) = 1$.

Finally we mention an application of the general relation (II.2.3.1) describing $\Pi_{i=1}^m L_S(1+is,\pi,\tilde{r}_i)$ in terms of $E(s,f,P,h)$. In Section (2.2), we mentioned that there exists f_v such that $W_v(e_v) \neq 0$. Consequently the zeros of $\Pi_{j=1}^m L_S(1+js,\pi,r_j)$ are among the poles of $E(s,f,e,P)$. From the fact that $E(s,f,e,P)$ is known to have no poles on the imaginary axis $i\mathbb{R}$ it follows that

$$\prod_{j=1}^{m} L_S(1+jit,\pi,r_j) \neq 0 \quad \forall t \in \mathbb{R} \ .$$

In particular, if we take $H = GL(n+m)$, and $G = GL_n \times GL_m$ (example (iii) on page 47 of [La1]), then we obtain exactly the non-vanishing theorem for the L-functions $L(s, \pi \times \pi')$ alluded to in Section (II.1.7).

In Chapter III it will be shown how to extend the partial L-functions $L_S(s, \pi, r_i)$ to include local factors at *every* place v, to prove an exact functional equation for these completed L-functions, and to prove that (in a wide variety of cases which include most of the L-functions heretofore discussed) these L-functions have only a finite number of poles.

Concluding Remarks.

(a) In 1983, C. Moreno circulated a 9-page manuscript sketching some of these ideas in special cases; cf. [Moreno2].

(b) As far as non-vanishing results for $L(s, \pi, r)$ on the line $Re(s) = 1$ are concerned, we note that such results in the case of the symmetric power L-functions $L(s, \pi_f, Sym^m(\rho_2))$, together with the *conjectured* holomorphy of all these functions for $Re(s) \geq 1$, would imply the truth of the so-called "Sato-Tate Conjecture" for π_f; cf. [Serre], pp.I.18-26, and [Sha9].

CHAPTER III.

RECENT DEVELOPMENTS (1982-)

In this Chapter we describe the most recent results available for the L-functions $L(s, \pi, r)$. Since all the L-functions considered here appear in the constant terms of Eisenstein series for some quasisplit reductive group, we can approach their analytic properties by either of the two methods discussed in previous Chapters.

§1. EXPLICIT CONSTRUCTION OF ZETA-FUNCTIONS Á LA PIATETSKI-SHAPIRO

In 1983, Piatetski-Shapiro and Rallis discovered a generalization of the method of Godement-Jacquet which makes it possible to treat the "standard" L-functions $L(s, \pi, \rho_{st})$ attached to any of the simple classical groups. Their method may also be regarded as a generalization of the method of Rankin-Selberg in that their global zeta-integrals involve Eisenstein series in the same way.

1.1. Origins of the method.

As alluded to already in Section II.(1.5), the origins of the theory of [PSR1] lie in the theory of dual reductive pairs and the Weil representation. We shall presently explain this in more detail following [Rallis3].

Fix G equal to the symplectic group of a $2n$ dimensional space V_1 equipped with a nondegenerate alternating form A, i.e., $G = Sp(n)$. Fix G' equal to the orthogonal group of an m dimensional quadratic space (V_2, Q). Then $Q \otimes A$ defines an alternating form on $V_2 \otimes V_1 = M_{m,2n}$ whose associated symplectic group $Sp(mn)$ naturally contains both G and G'. In fact G and G' mutually centralize each other in $Sp(mn)$ and hence comprise a "dual reductive pair" in the sense of [Howe]. The resulting "theta-correspondence" between automorphic forms on $G_{\mathbb{A}}$ and $G'_{\mathbb{A}}$ may be described as follows.

Let r denote the "Weil representation" of the (two-fold) metaplectic cover $\tilde{S}p(mn)$ of $Sp(mn)$ as given in [We2]; this representation acts in the Schwartz-Bruhat space of $M_{m,n}(\mathbb{A})$, and by restriction defines a representation of $\tilde{G}_{\mathbb{A}} \times G'$. For each $\Phi \in \mathcal{S}(M_{n,n}(\mathbb{A})$ we construct the "theta-function"

(III.1.1.1) $$\Theta_\Phi(x,y) = \sum_{\xi \in M_{m,n}(F)} (r(x,y)\Phi)(\xi) .$$

This is an automorphic form on $\tilde{G}_{\mathbb{A}} \times G'_{\mathbb{A}}$; when $n = m = 1$, and Φ is properly chosen, Θ_Φ essentially reduces to the classical theta series $\Theta(z)$ pulled up to (the metaplectic cover of) $Sp_1(\mathbb{A}) = SL_2(\mathbb{A})$; cf. Proposition 2.3.5 of [Ge4]. The "duality" or "theta" correspondence just alluded to is defined in terms of the above theta-kernel by the formula

(III.1.1.2) $$f \longmapsto F_f(g) = \int_{G_F \backslash G_{\mathbb{A}}} f(x)\Theta_\Phi(x,y)dx$$

where f is an automorphic cusp form on $\tilde{G}_{\mathbb{A}}$. Note that (III.1.1.2) essentially reduces to Waldspurger's correspondence II.1.5.2 when $n = 1$ and G' is the orthogonal group $PGL_2 = O(2,1)$ (the orthogonal group of the quadratic form of type (2,1)).

As in [Wald], a natural question to ask is this: *When is the lift $F_f(y)$ non-zero?* The answer lies in the computation of its norm:

$$\begin{aligned}
\|F_f\|^2 &= \int_{G(F)\backslash G(\mathbb{A})} F_f(y)\overline{F_f(y)}dy \\
&= \int_{G(F)\backslash G(\mathbb{A})} \int_{Sp(n)\times Sp(n)\backslash Sp(n)\times Sp(n)(\mathbb{A})} \\
&\quad \times \Theta_\Phi(x_1,y)f(x_1)\bar{\Theta}_\Phi(x_2,y)\overline{f(x_2)}dx_1dx_2dy \\
&= \int_{Sp(n)\times Sp(n)\backslash Sp(n)_{\mathbb{A}}\times Sp(n)_{\mathbb{A}}} f(x_1)\overline{f(x_2)} \\
&\quad \times \left\{ \int_{G(F)\backslash G(\mathbb{A})} \Theta_\Phi(x_1,y)\Theta_\Phi(x_2,y)dy \right\} dx_1dx_2 .
\end{aligned}$$

In [Rallis3], (a trick from) Kudla's work on dual reductive pairs (cf. [Kud]) is combined with the "Siegel Weil formula" of [We3] to interpret this last expression (in brackets) in terms of Eisenstein series! More precisely, $\|F_f\|^2$ is expressed as a special value of the meromorphic function

(III.1.1.3)

$$s \longmapsto \int_{G(F)\times G(F)\backslash G(\mathbb{A})\times G(\mathbb{A})} f(x_1)\overline{f(x_2)}E(s,G^*,(x_1,x_2))dx_1dx_2$$

where E denotes an Eisenstein series on some (bigger) symplectic group G^* which contains $G \times G$.

For the dual pair $(Sp(1), O(2,1))$, the function (III.1.1.3) can be expressed as the product of *two* terms. The first is a *global* one involving the (unramified part of) the Jacquet-Langlands L-function of the Shimura lift (§II.1.5) of the representation $\bar{\pi}$ containing f; the second is a *local* term which involves certain Whittaker type integrals for the components π_v of π, for only a finite number of "ramified" places v. The conclusion is that F_f is non-zero if and only if (i) the L-function $L_S(s_0, Shim(\bar{\pi}_f))$ does not vanish at a certain special value s_0 (this is the "global data") *and* (ii) certain components $\bar{\pi}_v$ possess a prescribed Whittaker model (the "local data"). This result is one of Waldspurger's strikingly original contributions, although his motivation and methods in [Wald] are different. The goal of [Rallis3] was to show that similar conclusions hold for higher dimensional dual reductive pairs.

We should remind the reader that the definition of the Weil representation (and hence the lifting II.1.1.2) depends on the choice of an additive character of $F\backslash\mathbb{A}$. (This is in fact what adds subtlety to the question of the non-vanishing of F_f.) We have ignored this point only because it is irrelevant to our subsequent discussion of L-functions. The crucial point for us is that the meromorphic function (II.1.1.3) can (in general) be related to higher rank L-functions (namely the "standard" L functions attached to $Sp(n)$), and this is our starting point for the discussion below.

1.2. The construction of Piatetski-Shapiro and Rallis.

Take G to be a reductive algebraic group with anisotropic center C (i.e. $C_F \backslash C_{\mathbb{A}}$ is compact). Let H be another reductive group over F, together with an embedding $G \times G \overset{i}{\hookrightarrow} H$, and let $G^d \subset H$ be the image of G under the diagonal embedding $G \overset{d}{\hookrightarrow} G \times G$ composed with i. Let P be a parabolic subgroup of H. Then $G \times G$ acts on the flag variety $X = P \backslash H$. We call a $G \times G$ orbit $X' \subset X$ *negligible* if the stabilizer R' (in $G \times G$) of a point $x' \in X'$ contains (as a normal subgroup) the unipotent radical N' of a proper parabolic subgroup of $G \times G$. Let $x_0 \in X$ be the point of X corresponding to the coset $P \cdot e$ and X_0 its $(G \times G)$-orbit; the stabilizer R_0 of x_0 in $G \times G$ is then just $P \cap (G \times G)$.

In order to construct a global zeta-integral related to cusp forms on $G_{\mathbb{A}}$, we need the following two conditions (on the $G \times G$ action on X) to be satisfied:

(1) The stabilizer R_0 of x_0 is G^d; and

(2) If X' is any orbit other than X_0, then X' is negligible. (Thus X_0 is called the main orbit; the justification for the term negligible will soon be clear from the construction below.)

In Section III.1.1 we prepared the reader for the appearance of the following Eisenstein series on H:

(III.1.2.1) $$E(\omega_s, f, h) = \sum_{P_F \backslash H_F} f_\omega(\gamma h) \ ,$$

Here $\omega_s : P_F \backslash P_{\mathbb{A}} \longrightarrow \mathbb{C}^x$ a quasicharacter which we shall assume is *trivial on* $G^d_{\mathbb{A}}$, and f_ω belongs to $ind_{P_{\mathbb{A}}}^{H_{\mathbb{A}}} \omega_s$. If π is an irreducible cuspidal representation of $G_{\mathbb{A}}$, and $\tilde{\pi}$ its contragredient, then to $\varphi_1 \in H_\pi$ and $\varphi_2 \in H_{\tilde{\pi}}$ we may associate the Rankin-Selberg type zeta-integral

(III.1.2.2)
$$\mathcal{Z}(\omega_s, \varphi_1, \varphi_2, f) = \int_{(G \times G)_F \backslash (G \times G)_{\mathbb{A}}} \varphi_1(g_1)\varphi_2(g_2)E(\omega_s, f, (g_1, g_2))dg_1 dg_2 \ .$$

Note that for $G = Sp(n)$, this expression essentially reduces to the integral (III.1.1.3) which arose in the theory of dual reductive pair liftings. In general, since φ_1 and φ_2 are cuspidal, this integral converges absolutely, and inherits the analytic properties of the Eisenstein series $E(\omega_s, f, h)$. It is precisely this global zeta-integral to which the local-global L-function machine must be applied.

Using the above assumptions on $P\backslash H/G \times G$ and ω_s, the "basic identity"

$$\text{(III.1.2.3)} \qquad \mathcal{Z}(\omega_s, \varphi_1, \varphi_2, f) = \int_{G_{\mathbb{A}}} f(g,1)\langle\pi(g)\varphi_1, \varphi_2\rangle dg$$

is established. The point is that the substitution of (1.2.1) into (1.2.2) leads to a sum over the double cosets $P_F\backslash H_F/(G\times G)_F$, and each of the "negligible orbit" contributions to $\mathcal{Z}(\omega_s, \varphi_1, \varphi_2, f)$ involves an expression of the form

$$\int_{N_F^i\backslash N_{\mathbb{A}}^i} \varphi_i(n_i m_i g_i) dn_i$$

where $i = 1$ or 2, and N^i is the unipotent radical of some parabolic $P_i \subset G$. Since φ_1 and φ_2 are cuspidal, this integral must vanish. Thus we have

$$\mathcal{Z}(s, \varphi_1, \varphi_2, f) = \Pi \mathcal{Z}((\omega_s)_v, \varphi_{1,v}, \varphi_{2,v}, f_v) \ .$$

In §2 of [PSR1] it is shown that for any classical group G of symplectic, orthogonal or unitary type, one can find H, P, and ω_s so that the conditions required for the basic identity above are satisfied. For example, if $G = Sp(n)$, then $H \approx Sp(2n)$.It remains to carry out steps (2)-(5) of Section I.1.7. Step (2) is carried out in §5 of [PSR1]. In particular, it is shown that $E(s, f, h)$ (and hence $\mathcal{Z}(s, \varphi_1, \varphi_2, f)$) has only *finitely many* poles in $\mathbb{C}$. Actually, the precise result is that $E(s, f, h)$ has only finitely many poles *after* it is multiplied, i.e. "normalized" by a suitable product of zeta-functions like $\zeta(2s)$. The need to normalize $E(s, f, h)$ should not seem surprising after our discussion of analogous results for $GL(2)$ in the Appendix to Section II(1.2).

Locally, it is also shown that the unramified zeta-integral at v, suitably normalized (by the local component of the global normalizing factor) is exactly equal to $L(s, \pi_v, r)$, where ${}^LG \longrightarrow GL_d(\mathbb{C})$ denotes the standard embedding of ${}^LG$; for example, if $G = Sp(n)$, then ${}^LG = SO_{2n+1}(\mathbb{C})$, and $r : {}^LG \longrightarrow GL_{2n+1}(\mathbb{C})$ is the standard embedding. The remaining local theory, including the functional equation and definition of local L and ε factors, is treated separately in [PSR2]. This theory is somewhat delicate, and we shall not attempt to explain it here. (See, however, the second paragraph in Section 1.3 below.) Instead we return to the global zeta-integral $\mathcal{Z}(s, \varphi_1, \varphi_2, f)$ and explain how it really does generalize the Godement-Jacquet integral when $G = GL(n)$.

For simplicity, we start with the group $G = PGL_n(F)$. Our first task is to construct a group H, a parabolic subgroup $P \subset H$, and an embedding $i : G \times G \hookrightarrow H$ satisfying conditions (1) and (2) above.To this end, let $W = M(n, F)$, a vector space of dimension $N = n^2$ over F. There is a natural action of $GL_n(F) \times GL_n(F)$ on W given by $x \cdot (g_1, g_2) = g_2^{-1} x g_1$, and the corresponding homomorphism $GL_n(F) \times GL_n(F) \longrightarrow GL_N(F)$ induces an embedding $i : G \times G \hookrightarrow PGL_N(F)$. Thus it is natural to take $H = PGL_N(F)$, and $i : G \times G \hookrightarrow H$ this embedding. Let e_0 in W correspond to the identity 1_n in $M(n, F)$, and take P to be the parabolic subgroup of H stablizing Fe_0. With this choice of data, we fix an automorphic cuspidal representation π of G, and ask how the zeta-integral $\mathcal{Z}(\omega_s, \varphi_1, \varphi_2, f)$ is related to the Godement-Jacquet $\mathcal{Z}$-function.

First we recall the Godement-Jacquet construction. With π as above (viewed now as a cuspidal representation of $GL_n \mathbb{A}_F$ with trivial central character), let $\xi(g)$ be the matrix coefficient $\xi(g) = \langle \pi(g)\varphi_1, \varphi_2 \rangle$, where $\varphi_1 \in H_\pi$ and $\varphi_2 \in H_{\tilde{\pi}}$. Then for Φ in $\mathcal{S}(M(n, \mathbb{A}_F))$, the Godement-Jacquet zeta-function may be written as

$$\mathcal{Z}(s, \xi, \phi) = \int_G |det\ g|^s \Phi(e_0 g) \xi(g) dg\ .$$

(Cf. (II.1.1.6), (II.1.1.7) in Section II.1.1.) On the other hand, this same $\Phi \in \mathcal{S}(M_n(\mathbb{A}))$ may be used to define an f in some induced representation from $P_{\mathbb{A}}$ by way of the formula

$$f^\Phi(h, s) = |det\ h|^s \int_{\mathbb{A}^x} \Phi(a e_0 h) |a|^{Ns} d^x a$$

and hence a Piatetski-Shapiro-Rallis zeta-integral

$$\mathcal{Z}(s,\varphi_1,\varphi_2,f^\Phi) = \int_{(G\times G)_F\backslash(G\times G)_{\mathbb{A}}} E(s,f^\Phi,(g_1,g_2))\varphi_1(g_1)\varphi_2(g_2)dg_1dg_2$$

(where $E(s,f^\Phi,h) = \Sigma_{P_F\backslash H_F} f^\Phi(\gamma g)$). A simple computation together with the basic identity (III.1.2.3) then implies that this last integral equals $\mathcal{Z}(ns,\xi,\Phi)$, as expected. See [PSR1] for details. We note that there are other possible (more natural) choices for H, P and i for GL_n, but they will not yield the Godement-Jacquet zeta-integrals.

1.3. Summing up.

The theory of Piatetski-Shapiro and Rallis generalizes the method of Godement-Jacquet from $GL(n)$ to the simple classical groups (of symplectic, orthogonal, and unitary type). In particular, it does not require the automorphic representations whose L-functions are analyzed to be *generic*. Even when $G = GL(n)$ (or more generally the unit group of a matrix algebra over a division algebra) it gives a new way of looking at the Godement-Jacquet zeta-function as a Rankin-Selberg type integral involving Eisenstein series on a much larger group.

As already mentioned, Steps (1)-(4) of the "L-function machine" of Section I.1.7 are carried out for the zeta-integrals of Piatetski-Shapiro and Rallis in [PSR1] and [PSR2]. A tentative solution to the problem (Step (5)) of defining L-functions at the ramified primes is suggested in [PSR2] by way of suitably normalized gamma factors for the local functional equation. More precisely, Whittaker functionals (defined with respect to *non*-maximal unipotent subgroups) are introduced in order to define some local coefficients as in [Sha3]; these local coefficients are then used to normalize the intertwining operators and gamma factors in the functional equation so that the resulting normalized objects satisfy equations of the type $M^* \circ M^* = I$. This procedure is similar to the approach to Jacquet's theory sketched in Appendix (A.2) of Section II.1.2. The new twist is that $L(s,\pi_v)$ can no longer be defined as the g.c.d. of certain zeta-integrals, but rather as the reciprocal of the numerator of the normalized gamma-factor (which is known to be a rational function in p^{-s}). As we shall see below, in Section III.2, the idea of defining local factors in this way appears in the recent work on local coefficients in [Sha8].More work on this local theory remains to be done before the poles of the (completed) L-function can be determined; cf., however, [PSR4].

1.4. Rankin triple products.

In 1985, an intriguing integral representation was discovered by P. Garrett for the L-function attached to a *triple* of classical modular forms, i.e., a Rankin-Selberg type representation for the product of *three* modular forms; cf. [Garrett1 and 2]. By embedding three copies of the upper half-plane H in the Siegel upper half space H_3 of degree 3, Garrett considered integrals of the form

(III.1.4.1)

$$L(\varphi_1, \varphi_2, \varphi_3, s) = \int \varphi_1(z_1)\varphi_2(z_2)\varphi_3(z_3)E((z_1, z_2, z_3), s)dz_1 dz_2 dz_3$$

with $E((\cdot), s)$ an Eisenstein on H_3 restricted to (a fundamental domain for) $H \times H \times H$; then he analyzed the Euler product expansion and meromorphic properties of this integral. Rather than describe this important classical construction in further detail, we turn instead to its group-theoretic formulation (and generalization) worked out in [PSR3].

Let K be a semi simple abelian algebra of degree 3 over a global field F, and let $GL_2(K)$denote the corresponding general linear group of degree 2. Note that in case $K = F \oplus F \oplus F$,

$$GL_2(K) = GL_2(F) \times GL_2(F) \times GL_2(F) \ .$$

On the space $V = K \oplus K$, consider the skew symmetric form given by $A[(x, y), (x', y')] = xy' - x'y$, and let $A' = tr_{K/F} \circ A$ denote the corresponding F-valued skew symmetric form on V. By $GS_p(A')$ we denote the group of similitudes of the form A', i.e., $GSp(A') = GSp_3(F)$. Moreover, we let $GL_2(K)^\circ = \{g \in GL_2(K) : \ g \in GSp_3(F)\}$; in particular, in case $K = F \oplus F \oplus F$, we have

$$GL_2(K)^\circ = \{(g_1, g_2, g_3) : \ g_i \in GL_2(F) \text{ and } det\ g_1 = det\ g_2 = det\ g_3\}$$

(corresponding to Garrett's embedding of $H \times H \times H$ in H_3). The Rankin integral generalizing (III.1.4.1) now takes the form

$$\text{(III.1.4.2)} \qquad \int_{Z'(\mathbb{A})GL_2(K)^\circ_F\backslash GL_2(K)^\circ(\mathbb{A})} F(x)E(x,s,f)dx = \mathcal{Z}(s,F,E)$$

with Z' the center of GSp_3.

In (III.1.4.2), F belongs to the space of an automorphic cuspidal representation Π of $GL_2(K)^\circ(\mathbb{A})$, and $E(s,x,f)$ is an Eisenstein series on $GSp_3(\mathbb{A})$ induced from (a quasicharacter ω_s of) a maximal parabolic subgroup P whose Levi component is isomorphic to $GL_3(F) \times F^x$. Note that when $K = F \oplus F \oplus F$, Π is a triple of automorphic cuspidal representations π_1, π_2, π_3 of $GL_2(\mathbb{A})$, and the integral (III.1.4.2) is essentially an adelization of (III.1.4.1). In general, there is a basic identity of the form

$$(III.1.4.3) \qquad \mathcal{Z}(s,F,E) = \int_{Z'(\mathbb{A})N(\mathbb{A})\backslash GL_2(K)^\circ(\mathbb{A})} f^\psi_{\omega_s}(g)W^\psi_F(g)dg$$

where W^ψ_F (resp. $f^\psi_{\omega_s}$) is a kind of Whittaker function attached to F in Π (resp. f_{ω_s} in $Ind_p^{GSp_3}\omega_s$) whose exact form depends on the nature of K. For example, if $K = F \oplus F \oplus F$, then $W^\psi_F(g_1,g_2,g_3) = \Pi^3_{i=1}W^\psi_{F_i}(g_i)$, with each $W^\psi_{F_i}$ the usual GL_2 Whittaker function.

Because $E(s,x,f)$ is defined by a summation over $P\backslash GSp_3$, whereas $F(x)E(s,x,F)$ is integrated over $GL_2(K)^\circ(F)\backslash GL_2(K)^\circ(\mathbb{A})$, it is not surprising that the proof of this basic identity involves a careful analysis of the (right) $GL_2(K)^\circ$ orbits in $P\backslash GSp_3$. In particular, it is shown that all but one of these orbits are negligible in the sense already explained in the case of [PSR1]. Once this basic identity is established, the idea is to once again apply the local global L-function machine of I.(1.7). In this case the most difficult step is perhaps the analysis of the meromorphic behavior of the Eisenstein series, i.e., the exact location of its poles. The result is a complete theory of L-functions attached to the following types of data:

(i) In case $K = F \oplus F \oplus F$, the group G in question is $GL_2 \times GL_2 \times GL_2$, and the representation r of its L-group $GL_2(\mathbb{C}) \times GL_2(\mathbb{C}) \times GL_2(\mathbb{C})$ is the tensor product $\rho_2 \otimes \rho_2 \otimes \rho_2$.

(ii) In case $K = F_2 \oplus F$, with F_2 a quadratic extension of F, we get $G = R_{F_2/F}(GL_2) \times GL_2$, where $R_{F_2/F}$ denotes the functor of restriction of scalars (from F_2 to F). Since $R_{F_2/F}(GL_2)(\mathbb{A}_F) \approx GL_2(\mathbb{A}_{F_2})$, Π is now of the form $\pi' \otimes \pi$, where π' is a cuspidal representation of $GL_2(\mathbb{A}_{F_2})$; $r = r_1 \otimes r_2$ is a certain eight dimensional representation of the twisted L-group ${}^LG$ (corresponding to the *non*-split group $R_{F_2/F}(GL_2) \times GL_2$ – see [PSR3] for details), where r_1 is a four dimensional representation irreducible representation of $(GL_2(\mathbb{C}) \times GL_2(\mathbb{C})) \times Gal(F_2/F)$, the L-group of $R_{F_2/F}(GL_2)$, to which Asai attached the L-function $L(s, \pi', r_1)$. (Cf. [Asai], [HLR], and especially the remarks at the end of Section II.1.8 above.)

(iii) In case $K = F_3$, a cubic extension of F, we have $G = R_{F_3/F}(GL_2)$ and $G_{\mathbb{A}_F} = GL_2(\mathbb{A}_{F_3})$; thus Π corresponds to an automorphic cuspidal representation of $GL_2(\mathbb{A}_{F_3})$. This generalizes Asai's L-function discussed in (ii) from quadratic extensions to cubic ones.

We note that in [PSR3], the local theory of L-functions at the ramified places is developed as in the Appendix (A.2) to Section II.1.2, and as in the theory for $G \times GL(n)$ sketched below. In particular, a certain Eisenstein series is shown to be "nice" only for functions in the induced space of type (i), (ii) or (iii) as in the Appendix A.2 of Section II.1.2, and the local L-factors are defined as g.c.d.'s of the family of local zeta-integrals with f again of type (i), (ii) or (iii).

We refer to Section III.2.1 below for a different approach to these results, following the methods of [Sha8].

We conclude this Section with some remarks about the significance of "triple products" and their possible generalization. Given π_1, π_2 on GL_2, the fact that $L(s, \pi_1 \times \pi_2 \times \pi_3)$ is "nice" for each π_3 should imply (by the "converse theorem for GL_4") the existence of a Π on $GL(4)$ such that $L(s, \Pi \times \pi_3) = L(s, \pi_1 \times \pi_2 \times \pi_3)$ (the L-function on the left side being a Rankin-Selberg convolution on $GL_4 \times GL_2$); repeating this procedure for a second pair π_3, π_4, with lifting Π' on GL_4, should produce a "quadruple" L-function, namely $L(s, \pi_1 \times \pi_2 \times \pi_3 \times \pi_4) = L(s, \Pi \times \Pi')$. Since triple products are not yet understood for $GL(3)$ or $GL(4)$, this seems to be as far as we can go with multiple products on GL_2 (namely 4).

Finally, we remark that the converse theorem for GL_4 should – together with result (ii) above– imply that there exists an automorphic representation of $GL_4(\mathbb{A}_F)$ whose standard L-function equals Asai's L-function $L(s,\pi',r_1)$. Unfortunately, a detailed proof of the converse theorem for GL_4 has not yet appeared in print.

1.5. L-functions for $G \times GL(n)$.

In [GePS5], Gelbart and Piatetski-Shapiro develop several new generalizations of the Rankin-Selberg method which make it possible to prove Langlands' Conjecture for non-simple groups of the form $G \times GL(n)$, at least when G has split rank n. For these L-functions, the proof of the conjecture is ultimately related (via the "converse theorem") to the lifting of automorphic forms between G and an appropriate $GL(m)$. A special example of this phenomenon already appears in the work of [PSS1, 2] on $GSp_4 \times GL_2$, where the relevant lift is from GSp_4 to GL_4, and in the work of [GePS1] on $U_{2,1} \times GL_1(K)$, where the lifting is "base change" from $U_{2,1}$ to $GL_3(K)$ (K the quadratic extension of k defining the unitary group $U_{2,1}$); cf. §II.1.8(c).

A prototype of the Rankin-Selberg approach used in [GePS5] may be described as follows. Suppose G is the split special orthogonal group SO_{2n+1} over F, and H denotes the subgroup of G isomoprhic to $SO(2n)$ (the split special orthogonal group in $2n$ variables). Then H has a maximal parabolic subgroup P whose Levi component is isomorphic to $GL(n)$. Given automorphic cuspidal representations π and τ, of G and $GL(n)$ respectively, we form the integral

$$I(s,\varphi,E) - \int_{H_F\backslash H_{\mathbb{A}}} \varphi(h)E_\tau^f(h,s)dh \tag{III.1.5.1}$$

where φ belongs to the space of π, and E_τ^f is an Eisenstein series on H associated to τ, i.e., induced from the cuspidal representation $\tau \otimes |det|^{s'} \otimes 1$

of P. We note that the constant term of this Eisenstein series involves the automorphic L-function $L(2s, \tau, \Lambda^2)$, where Λ^2 denotes the exterior (or antisymmetric) square of the standard representation of $GL_n(\mathbb{C})$; more precisely, it is $L_S(2s, \tau, \Lambda^2)$ which appears in the analogue of (I.2.5.3) for the present Eisenstein series.

Assuming that both π and τ possess Whittaker models, we can establish the "basic identity"

$$I(s, \varphi, E^f) = \int_{U^H_{\mathbb{A}} \backslash H_{\mathbb{A}}} W(h) W_\tau(h, s) dh$$

where U^H is the standard maximal unipotent of H, $W(h)$ belongs to the Whittaker model of π, and $W_\tau(h, s)$ belongs to the space of $ind\ \tau \otimes |det|^{s'}$. As expected, the right hand side of this identity factors as a product of similar local zeta-integrals, one for each place of F.

In [GePS5] the groundwork is laid for a local theory of these zeta-integrals along the lines sketched for $GL(2) \times GL(2)$ in the Appendix A.2 to Section II.1.2. To explain this, consider the local zeta-integral

$$\mathcal{Z}(s, W, f_\tau) = \int_{U^H \backslash H} W_\pi(h) f_\tau(h, s) dh ,$$

where π is an irreducible admissible representation of $G = SO_{2n+1}$ over the local field F, $W_\pi \in \mathcal{W}(\pi, \psi)$, and $f_\tau(h, s)$ inside $ind\ \tau \otimes |det|^{s'}$ is such that $f_\tau(mh, s) = w_\tau(m)$ belongs to $\mathcal{W}(\tau \otimes |\ |^{s' + \frac{n-1}{2}}, \psi^{-1})$ for each h in H.

As in the Appendix A.2 to Section II.1.2, we study this family of zeta-integrals $\mathcal{Z}(s, W, f_\tau)$ *only* as $f_\tau(h, s)$runs through "good sections" of $\mathbb{H}(s, \tau) = Ind\ \tau \otimes |\ |^{s'}$. Thus we suppose that the following conditions are satisfied:

(III.1.5.2) Either (i) τ is unramified, and $f_\tau(k, s) \equiv L(2s, \tau, \Lambda^2)$ for all k in the maximal compact subgroup K^H of H; or (ii) the restriction of f_τ to K^H is independent of s; or (iii) $f_\tau = A^*(-s', \tilde{\tau}) f^*_\tau$, where $A^*(-s', \tilde{\tau})$ is a (normalized) operator intertwining $Ind\ \tilde{\tau} \otimes |\ |^{-s'}$ with $Ind\ \tau \otimes |\ |^{s'}$, and f_τ^* is of type (ii).

Applying Iwasawa's decomposition $H = U^H A K^H$, we conclude (as in II.1.2) that $\mathcal{Z}(s, W, f)$ is a (finite) sum of integrals of the form

$$f_\tau(k_0, s) \int_A W_\pi(a) W_\tau(a) |det\ a|^{s' + \frac{n-1}{2}} d^x a$$

where $k_0 \in K^H$, $W_\pi \in \mathcal{W}(\pi, \psi)$, $W_\tau \in \mathcal{W}(\tau, \psi^{-1})$, and A denotes the maximal split torus of H. Note A may also be regarded simultaneously as the maximal torus of G (namely $\{diag(a_1, \ldots, a_n, 1, a_n^{-1}, \ldots, a_i^{-1}) : a_i \in F^x\}$) or $GL_n(F)$ (namely $\{diag(a_1, \ldots, a_n) : a_i \in F^x\}$). To analyze the arithmetic of these integrals, we appeal directly to the theorem of [CS] describing the asymptotic behavior of arbitrary Whittaker functions (cf. Appendix A.2 to Section II.1.2 for $GL(2)$).

We are assuming now that our local field is nonarchimedean. In the present context, Proposition 6.3 of [CS] says that for $(a_1, \ldots, a_n)$ *near zero*, $W_\pi(a) W_\tau(a)$ behaves like a finite linear combination of functions of the form $\lambda : (a_1, \ldots, a_n) \longmapsto \Pi_{i=1}^m \mu_i(a_i) v(a_i)^{m_i}$, where μ_i is a quasicharacter of F^x, v is the discrete valuation of F, and m_i is a nonnegative integer. In fact, the collection of such λ needed to describe the behavior of any $W_\pi\ \ W_\tau$ near zero is finite, and depends only on π and τ. Since it is an elementary fact (cf. Prop. 6.1 of [CS]) that $W_\pi W_\tau \equiv 0$ for large a_i, it follows that the integrals appearing in (III.1.5.2) are rational functions (in q^{-s}) with a common denominator.

In Chapter II of [GePS5] it is shown (by the theory of $GL(n)$ representations developed in [GelKaz] but without appealing to [CS]) that the zeta-integrals $\mathcal{Z}(s, W_\pi, f)$ are meromorphic, and satisfy the functional equation

$$\mathcal{Z}(1 \quad s, W, M(s)f) = \gamma(s, \pi \times \tau) \mathcal{Z}(s, W, f) .$$

From this, and the above discussion, it follows that the family of local zeta-integrals $\mathcal{Z}(s, W, f)$ generates a fractional ideal in $\mathbb{C}[q^s, q^{-s}]$. From the fact that $\mathcal{Z}(s, W, f) \equiv 1$ for *some* W and f (with f of type (ii); cf.

Prop. 12.4 of [GePS5])it follows (as for $GL(2)$) that there is a (unique) generator $L(s, \pi \times \tau)$ of this ideal of the form $\frac{1}{Q_0(q^{-s})}$ with $Q_0(0) = 1$. Thus, as in the Appendix A.2 to Section II.1.2, we can also define ε-factors $\varepsilon(s, \pi \times \tau)$ and go on to complete the local and global theory of L-functions. In particular, we can identify each "unramified" zeta-integral with the Langlands local factor $L(s, \Pi_v, r_v)$, where $\Pi_v = (\pi_v, \tau_v)$, and $r : Sp_{2n}(\mathbb{C}) \times GL_n(\mathbb{C}) \longrightarrow GL_{2n^2}(\mathbb{C})$ is the standard tensor imbedding. We note here that the factor $f_\tau(k_0, s) = L(2s, \tau, \Lambda^2)$ in (III.1.5.2) plays a crucial role in producing the correct local factor (cf. the Appendix to [GePS5]); this is the local scalar factor which appears so magically in Jacquet's $GL(2) \times GL(2)$ theory.

Globally, the goal is to analyze the analytic continuation of $E_\tau^f(h, s)$ *for good sections* $f = \Pi f_v$ locally satisfying the conditions (i)-(iii) just discussed. Of course, f_v is almost everywhere "unramified" and satisfies condition (i). As discussed in more detail in the Appendix A.2 of Section II.1.2 for the special case of $GL(2) \times GL(2)$, the idea is that the "global normalizing factor" $L_S(2s, \tau, \Lambda^2)$ knocks out the infinitely many unwanted poles of $E(h, s)$.

A detailed proof of this fact will involve proving that an analogue of Lemma A.2.4 of the Appendix A.2 holds in the present general context – with $L(2s, \chi_p)$ replaced by the local factor of $L(2s, \tau, \Lambda^2)$. Since these latter L-functions are now known to have only finitely many poles (cf. the discussion of Shahidi's recent work in §2.5 below), the implication is that $L(s, \pi \times \tau)$ itself has only finitely many poles.

We note that for $G = SO_{2n}$, the form of the analogous zeta-integral looks similar, but the actual derivation of the identity requires a different method. Indeed in this case, inspired directly by the work of [Garrett1,2] and [PSR3], the roles of G and H are reversed: the basic integral is taken *over* G, while the Eisenstein series is defined on $H = SO_{2n+1}$ and restricted to G. On the other hand, for the case of $G = Sp_{2n}$, even the form of the integral for $I(s, \varphi, E)$ is new, requiring as it does a theta-series in the integral; in this case, no subgroup appears at all, and the form of the integral is directly modeled on (a generalization of) Shimura's method. In both cases, the lower rank L-function which plays the role of $L(s, \tau, \Lambda^2)$ for $G = SO_{2n+1}$ is $L(s, \tau, Sym^2)$.

§2. LANGLANDS' THEORY COMPLETED

As already alluded to many times in this work, all the L-functions $L(s,\pi,r)$ thus far studied appear in a constant term of some Eisenstein series on a quasi-split reductive group. Indeed they were first introduced in this way by Langlands (at least for split groups) and through them he was led to make many of the deep conjectures of [La3] (including Conjecture L on the analytic properties of $L(s,\pi,r)$).

Our purpose now is to survey the theory of [Sha8] which gives the most general results on $L(s,\pi,r)$ to date. The first type of general result in [Sha8] concerns a uniform line of absolute convergence for the Euler products defining $L(s,\pi,r)$; this result is also related to a uniform estimate for the Hecke eigenvalues of (generic) cusp forms on many quasi-split groups, and in particular gives the best available estimates for the Fourier coefficients of Maass wave forms. A second type of result concerns not only the functional equation and meromorphic continuation of each $L(s,\pi,r)$ but also a proof that $L(s,\pi,r)$ has only *finitely* many poles in $\mathbb{C}$ for a wide and important subclass of $L(s,\pi,r)$'s which includes not only the L-functions studied by Piatetski-Shapiro and Rallis (at least for generic representations), but also the third symmetric power of L-functions attached to cusp forms on PGL_2.

2.1. Range of applicability of the method.

As in §II.(2.1), we are dealing with ψ*-generic* automorphic cuspidal representations $\pi=\otimes\pi_v$ of a reductive group G, where $\psi_U=\Pi\psi_v$ is a generic character of a maximal unipotent subgroup U of the reductive group H (containing G as the Levi component of one of its maximal parabolic subgroups P). For such π, recall also that we have the following relations

$$\text{(III.2.1.1)}\quad M(s,\pi)f=\left(\bigotimes_{v\in S}A(s,\pi_v)f_v\right)\otimes\bigotimes_{v\notin S}\tilde{f}_v^0\prod_{i=1}^{m}\frac{L_S(is,\pi,\tilde{r}_i)}{L_S(1+is,\pi,\tilde{r}_i)},$$

$$E_\psi(s,f,P,e) = \left(\prod_{v\in S} W_v(e_v)\right)\prod_{i=1}^{m} L_S(1+is,\pi,\tilde r_i)^{-1} \tag{III.2.1.2}$$

and the functional equation

$$\prod_{i=1}^{m} L_S(is,\pi,r_i) = \left(\prod_{v\in S} C_{\bar\psi_v}(s,\tilde\pi_v)\right)\prod_{i=1}^{m} L_S(1-is,\pi,\tilde r_i)\,. \tag{III.2.1.3}$$

These were some of the main results of [Sha3] explained in Sections II.(2.1)-(2.4) of the present work. See formulas (I.2.5.3), (II.2.3.1) and (II.2.4.3) especially; recall that we have normalized matters so that the eigenvalues a_i for each of the representations $r_i : {}^LG \longrightarrow GL({}^L\mathcal{N})$ are $i = 1,\ldots,m$.

In Section 4 of [Sha8], an exhaustive study is made of the representations which might possibly arise as r_i's in the above context. This involves first of all completing the list (or tables) in the case of split groups ([La1], pp. 46-53), and secondly treating the quasi-split case. In all these lists, the highest weight λ_i of r_i is written as a linear combination of the fundamental weights δ_j. It is interesting and instructive to single out more particular examples from these lists.

From page 51 of [La1] consider the example

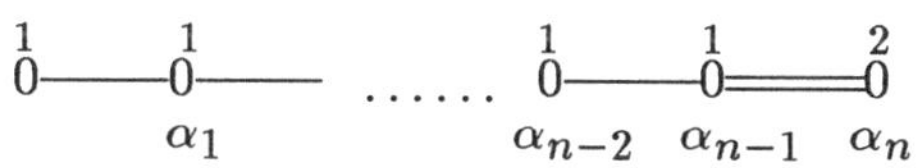

(xx) $\qquad m = 1 \qquad \lambda_1 = \delta_1 \qquad a_1 = 2(n+1)$

The full Dynkin diagram tells us that we are dealing with the group $H = Sp(n+1)$; the diagram corresponding only to the labelled vertices $\alpha_1,\ldots,\alpha_n$ is still of type C_n, but now of rank n, i.e. $G = Sp(n)$. Since the highest weight of r_1 is just $(1,0,\ldots,0)$, this means that we are dealing with the standard embedding $r_1 : {}^LG \longrightarrow GL_{2n+1}(\mathbb{C})$, where ${}^LG = SO_{2n+1}(\mathbb{C})$, and the resulting L-function $L(s,\pi,r_1)$ is just the $Sp(n)$ L-function treated in [PSR1]. The other L-functions of [PSR1] arise in examples (xvii) ($SO(2n+1)$) and (xxiii) ($SO(2n)$) of [La1]. The standard Godement-Jacquet L-function for $GL(n)$ arises from example (iii), page 47.

We take this opportunity to point out that on page 50 of [La1], in case (xvii), the roles of λ_1 and λ_2 should be interchanged, i.e., $\lambda_1 = \delta_1$ while $\lambda_2 = 0$; the same is true for example (xxi).

The L-functions for "$G \times GL(n)$" of [GPS5] may be read off only from the tables in [Sha8] which complete Langlands' list. For example, $SO(2k+1) \times GL_k$ appears as example (B_n) of that paper. Here *two* L-functions $L(s, \pi, r_i)$ appear in the constant term of $E(s, h)$: the first has highest weight $\delta_1 + \delta_{k+2}$, which means that $r : \; Sp_k(\mathbb{C}) \times GL_k(\mathbb{C}) \longrightarrow GL_{2k^2}(\mathbb{C})$ corresponds to the standard tensor product of the standard representations of $Sp_k(\mathbb{C})$ and $GL_k(\mathbb{C})$; the second has highest weight $2\delta_1$ (and in particular, is not one-dimensional). The case of $SO(2k) \times GL(k)$ (resp. $Sp(k) \times GL(k)$) corresponds to the example labelled $D_n - 1$ (resp. C_n) in this same paper; as we mentioned earlier (in §II.2.6), the L-functions for $GL(n) \times GL(m)$ arise from example (iii) of Langlands' table. In all these examples, the reader should take care to note that the role of the Levi component $G \subset P$ is played by what we have been denoting $G \times GL(n)$ in the Rankin-Selberg theory.

Before describing the results on $L(s, \pi, r_i)$ obtained in [Sha8], we discuss one last example from the list of possible r_i's. In example $D_n - 2$ of Section 4 of [Sha8], let $n = 4$, so that H may be a split adjoint group of type D_4 $\left(\begin{array}{ccc} & & \underset{\alpha_3}{0} \\ \underset{\alpha_1}{0} & \!\!—\!\! & 0 \\ & & \underset{\alpha_4}{0} \end{array} \right)$, and $G \approx GL_2 \times SO(2,2)$ (with a surjection onto $GL_2 \times PGL_2 \times PGL_2$) is the Levi component of the parabolic subgroup belonging to the (disconnected) Dynkin diagram $\underset{\alpha_1}{0} \quad \begin{array}{c} 0\alpha_3 \\ 0\alpha_4 \end{array}$. Then $L(s, \pi, r_1)$ is exactly the Rankin triple product $L(s, \pi_1 \times \pi_2 \times \pi_3, \rho_2 \otimes \rho_2 \otimes \rho_2)$ considered in [Garrett] and [PSR3].On the other hand, the "twisted" examples discussed in §III.1.4, corresponding to quadratic and cubic extensions, belong to cases ${}^2A_3 - 2$ (Asai's L-function) and ${}^3D_4 - 1$ in Section

4 of [Sha8]. (More generally, the GL_n "quadratic" L-functions discussed in [Flicker1] correspond to examples ${}^2A_{2n+1} - 2$ in [Sha8].)

It is, hopefully, clear now how the theory to be described below includes almost all the automorphic L-functions previously studied.

2.2. A uniform line of convergence for $L_S(s, \pi, r)$.

Recall that $L_S(s, \pi, r)$ refers to the (partial) Langlands L-function for the data (G, π, r); it is defined by the infinite Euler product

$$\text{(III.2.2.1)} \qquad \prod_{v \notin S} det[I - r_v(A_{\pi_v})q_v^{-s}]^{-1}$$

where S denotes any finite set of places v outside of which π_v (and also χ_v and G_v) is "unramified". Theorem 5.1 of [Sha8] asserts that this infinite product always converges in the half-plane $Re(s) > 2$. As we shall explain below this fact is closely related to the holomorphy of the local intertwining operators $A(s, \pi_v)$ in the half-plane $Re(s) \geq 1$ (Thm. 5.2 of [Sha8]).

Let us *assume* first that for each *finite* v in S, the intertwining operator $A(s, \pi_v)$ is holomorphic for $Re(s) \geq 1$. In order to prove that the Euler product (III.2.2.1) converges absolutely for $Re(s) > 2$, it suffices to prove that each unramified factor

$$L(s, \pi_v, r) = det(1 - r_v(A_{\pi_v})q_v^{-s})^{-1} = \prod_{i}^{deg\ r} (1 - \alpha_{i,v}q_v^{-s})^{-1}$$

is holomorphic for $Re(s) \geq 1$. Indeed, this latter holomorphy implies the non-vanishing of each factor $1 - \alpha_{i,v}q_v^{-s}$, where q_v (some power of *a* prime) is the order of the residue class field of F_v; in particular

$$\text{(III.2.2.2)} \qquad |\alpha_{i,v}| < q_v$$

and this gives the required convergence. In Section 2.3 below we shall explain the connection between (III.2.2.2) and more familiar Ramanujan-type estimates. First we sketch the proof of (III.2.2.2).

In order to prove that each unramified $L(s, \pi_v, \tilde{r})$ is holomorphic for $Re(s) \geq 1$, we simply enlarge S (if necessary) to include v. Then (by assumption) $A(s, \pi_v)$ is holomorphic for $Re(s) \geq 1$. For simplicity of exposition, we shall assume $m = 1$. Because v is unramified,

$$A(s, \pi_v) f_v^0 = \frac{L(s, \pi_v, \tilde{r})}{L(1 + s, \pi_v, \tilde{r})} \tilde{f}_v^0 ,$$

with f_v^0 the unique normalized K_v^H-fixed function in $I(s, \pi_v)$. Thus we know that at least this *quotient* of unramified L-factors is holomorphic for $Re(s) \geq 1$. Now pick $N_0 \geq 1$ so large that $L(1+s, \pi_v, \tilde{r})$ (the reciprocal of a polynomial in q^{-s}) has no poles for $Re(s) \geq N_0$. Then $L(s, \pi_v, \tilde{r}) = L(1+s, \pi_v, \tilde{r}) \frac{L(s,\pi_v,\tilde{r})}{L(1+s,\pi_v,\tilde{r})}$ has no poles for $Re(s) \geq N_0$, and arguing inductively gives us the conclusion that $L(s, \pi_v, \tilde{r})$ is holomorphic for $Re(s) \geq 1$. (For $m > 1$, a technical inductive lemma – Lemma 4.2 of [Sha8] – is needed.)

It remains to prove our assumption, namely that for each finite $v \in S$, $A(s, \pi_v)$ is holomorphic for $Re(s) \geq 1$.

Step 1. Observe that for $Re(s) \geq 0$, the product

$$\prod_{i=1}^{m} L_S(1 + is, \pi, \tilde{r}_i)$$

has only a finite number of simple zeros.

Indeed relation (III.2.1.2) implies that the poles of $\Pi_{i=1}^m L_S(1 + is, \pi, \tilde{r}_i)^{-1}$ are the same as those of $E_\chi(s, f, P, e)$ (choosing W_v so that $W_v(e) \neq 0$ for $v \in S$), and the poles of this latter function in the right half-plane are known to be (finitely many) simple ones.

Step 2. Prove that for $Re(s) \geq 1$, both $L_S(s, \pi, \tilde{r}_i)$ and $\frac{L_S(s,\pi,\tilde{r}_i)}{L_s(1+s,\pi,\tilde{r}_i)}$ have only a finite number of poles and zeros, $1 \leq i \leq m$.

For simplicity of exposition we shall (once again) assume that $m = 1$. From relation (III.2.1.1) we may conclude that $\frac{L_S(s,\pi,\tilde{r}_i)}{L_S(1+s,\pi,\tilde{r}_i)}$ has only finitely many poles in $Re(s) \geq 1$; indeed for each v in S (and each s), we can choose f_v so that $A(s, \pi_v) f_v \not\equiv 0$. Picking N_0 so large that $L_S(s, \pi, r)$

has no poles (in particular, picking s_0 to the right of the line of absolute convergence for the Euler product $L_S(s,\pi,\tilde{r})$ established in [La1]) we may argue inductively as before to conclude that

$$L_S(s,\pi,\tilde{r}) = L_S(1+s,\pi,\tilde{r})\frac{L_S(s,\pi,\tilde{r})}{L_S(1+s,\pi,\tilde{r})}$$

also has only finitely many poles for $Re(s) \geq 1$. Since the zeros of the quotient $\frac{L_S(s,\pi,\tilde{r})}{L_S(1+s,\pi,\tilde{r})}$ come either from the poles of the denominator or the zeros of the numerator, what we have proved so far, including Step 1, shows that this quotient too only has finitely many zeros.

Step 3. Conclude that $\otimes_{v\in S} A(s,\pi_v)f_v$ has no poles in $Re(s) \geq 1$.

This follows almost directly now from relation (III.2.1.1) describing $M(s,\pi)f$ in terms of $\frac{L_S(s,\pi,\tilde{r})}{L_S(1+s,\pi,\tilde{r})}$. Indeed fix v finite but otherwise arbitrary in S, and for each $u \neq v$ in S pick f_u so that $A(s,\pi_u)f_u(e)$ is a constant independent of s if u is finite. Now suppose for some s with $Re(s) \geq 1$, $A(s,\pi_v)$ has a pole. For each $u|\infty$ pick f_u such that $A(s,\pi_u)f_u \neq 0$. Since $A(s,\pi_v)f_v$ is a rational function in q_v^{-s}, it will then have infinitely many poles parallel to the imaginary axis. Consequently, by Step 2 above,

$$M(s,\pi)f = \left(\bigotimes_{v\in S} A(s,\pi_v)f_v\right)\frac{L_S(s,\pi,\tilde{r})}{L_S(1+s,\pi,\tilde{r})}\bigotimes_{v\notin S}\tilde{f}_v^0$$

must also have infinitely many poles in $Re(s) \geq 1$. This contradiction completes the proof of the fact that for each finite $v \in S$, $A(s,\pi_v)$ is holomorphic for $Re(s) \geq 1$, hence also that each *unramified* $L(s,\pi_v,r)$ is holomorphic in this same half-plane.

2.3. Ramanujan-type estimates.

The eigenvalue estimate (III.2.2.2) is strictly weaker than what one expects from the "generalied Ramanujan Conjecture", since the latter Conjecture requires that

$$(*) \qquad |\alpha_{i,v}| \leq 1\,.$$

Similarly, the uniform line of convergence established above for the Euler product defining $L_S(s,\pi,r)$ is not as good as what would follow from Ramanujan-type estimates like $(*)$. In particular, one expects (for any *generic* cuspidal π) that this Euler product converges for $Re(s) > 1$, i.e., up to the line where it is known to be non-vanishing. For $G = GL_n \times GL_m$, such a result is proved, even though the stronger estimate $(*)$ is not yet established (see [JS1]). In general, the fact that

$$L_S(s,\pi,r) = \prod_{v \notin S'} L(s,\pi_v,r)$$

converges absolutely for $Re(s) > 2$ is the best result available to date, and it already has some interesting corollaries.

First of all, note that when (III.2.2.2) is applied in the context of $GL(n) \times GL(n)$ embedded as the Levi component of the (n,n) parabolic subgroup of $GL(2n)$ (example (iii) on p.47 of [La1]), the conclusion is that

$$|\mu| < q_v^{\frac{1}{2}}$$

for each eigenvalue μ of the conjugacy class A_v attached to an unramified irreducible representation $\sigma_v$of $GL_n(F_v)$. This is precisely Cor.2.5 of [JS1]. It is obtained by applying (III.2.2.2) to the cuspidal representation $\pi = \sigma \otimes \sigma$ of $M_{\mathbb{A}} = GL_n(\mathbb{A}) \times GL_n(\mathbb{A})$.

Now suppose π is any *non-monomial* cuspidal representation of $PGL_2(\mathbb{A}_F)$. (For our purposes here, "non-monomial" means that the Gelbart-Jacquet lift of π to $PGL_3(\mathbb{A})$ is again cuspidal.) At each unramified place v, let $A_v = \begin{bmatrix} \alpha_v & \\ & \alpha_v^{-1} \end{bmatrix}$ be the corresponding semi-simple conjugacy class in $SL_2(\mathbb{C})$; then

$$q_v^{-1/5} < |\alpha_v| < q_v^{1/5} \ .$$

This is the best available estimate for the Hecke eigenvalues of Maass wave forms (cf. [Moreno-Shahidi2] where the weaker estimate $p^{-1/5} \leq |\alpha_p| \leq p^{1/5}$ is proved only when $F = \mathbb{Q}$). The new estimate can be proved by applying (III.2.2.2) to a split group of type F_4 (example ($F_4 - 2$) in §4 of [Sha8], with G generated by $\{\alpha_1, \alpha_2, \alpha_4\}$) and the representation $\Pi \otimes \pi$ of G, where Π is the Gelbart-Jacquet lifting of π to $PGL_3(\mathbb{A})$.

2.4. Analytic continuation of the completed L-function.

The second half of [Sha8] extends the partial L-functions $L_S(s, \pi, \tilde{r}_i)$ to include local L (and ε) factors at every place, and proves that these completed L-functions have only finitely many poles in $\mathbb{C}$ (as well as a standard functional equation). These results are established *only* if the pair (H, G) corresponding to $L(s, \pi, \tilde{r}_i)$ is such that $m = 1$ or 2, and r_2 is at most one-dimensional. This includes almost all the interesting cases thus far discussed, but *not* the $G \times GL(n)$ theory of [GePS5] (cf. the discussion of Shahidi's tables in §2.1 above).

For simplicity of exposition, we shall once again assume that H splits over F, and $m = 1$. For any archimedean place v, there is a "Langlands parameter"

$$\varphi_v : W_{F_v} \longrightarrow {}^L G$$

associated to the irreducible representation π_v of the real or complex group G_v. (This is the "Galois" classification of irreducible representations alluded to as a Conjecture during our discussion of L-groups in Section I.2.3; for archimedean places, it is a theorem proved in [La6]. There is also a corresponding local Artin L-function $L(s, r \cdot \varphi_v)$ (cf. [La7]), where $r \cdot \varphi_v$ is now a representation of the Weil group W_{F_v}. Thus we may extend $L_S(s, \pi, r)$ across the archimedean places by way of the formula

$$L_S(s, \pi, r) = \prod_{v=\infty} L(s, r \cdot \varphi_v) \prod_{\substack{v \notin S \\ v < \infty}} L(s, \pi_v, r) \; .$$

Recall once again the striking similarity between the functional equation II.2.4.3 for $L_S(s, \pi, r)$ and the functional equation I.1.5.3 for Jacquet-Langlands' $L(s, \pi)$. This, together with the work of [Sha4], which relates the archimedean "local coefficients" $C_{\psi_v}(s, \pi_v)$ to Artin factors, thus leads one to define the local "gamma factors"

$$\gamma(s,\pi_v,r,\bar{\psi}_v) = C_{\bar{\psi}_v}(s,\tilde{\pi}_v)$$

for each $v \in S \cup \{\infty\}$. (Now, and henceforth, we let S denote the set of all *finite* ramified places.) A double application of the functional equation (III.2.1.3) also yields the expected identities

$$\prod_{v \in S \cup \{\infty\}} \gamma(s,\pi_v,r,\bar{\psi}_v)\gamma(1-s,\pi_v,\tilde{r},\psi_v) = 1 \tag{III.2.4.2}$$

and (by Theorem 3.1 of [Sha4])

$$\prod_{v \in S} \gamma(s,\pi_v,r,\bar{\psi}_v)\gamma(1-s,\pi_v,\tilde{r},\psi_v) = 1\ . \tag{III.2.4.3}$$

We now extend S as follows. For every v in S, add to S all those places of F which lie over the same rational prime (of $\mathbb{Q}$) as v does. To define $L(s,\pi_v,r_v)$ for any $v \in S$ we proceed as follows. Because $C_{\psi_v}(s,\pi_v)$ (and hence $\gamma(s,\pi_v,r,\bar{\psi}_v)$) is a rational function of q_v^{-s}, we may choose a polynomial P_v in $\mathbb{C}[q^{-s}]$ such that $\Pi_{v\in S}P_v(q_v^{-s})$ is the numerator of $\gamma_S(s,\pi,r,\bar{\psi}) = \Pi_{v\in S}\gamma(s,\pi_v,r,\bar{\psi}_v)$, and $P_v(0) = 1$. Similarly we choose $\tilde{P}_v$ associated to $\gamma_S(s,\pi,\tilde{r},\psi)$. Then from (III.2.4.3) it follows that

$$\gamma_S(s,\pi,r,\bar{\psi}) = \varepsilon_S(s,\pi,r,\psi)\prod_{v\in S}\frac{\tilde{P}_v(q_v^{-(1-s)})^{-1}}{P_v(q_v^{-s})^{-1}} \tag{III.2.4.4}$$

with ε_S is a product of monomials in q_v^{-s}, $v \in S$. Moreover, from the full functional equation it follows that P_v and $\tilde{P}_v$ are independent of ψ (as long as all components outside S are unramified). Thus it is natural to extend $L_S(s,\pi_v,r)$ across all places through the formulas

$$L(s,\pi_v,r) = P_v(q_v^{-s})^{-1} \quad \text{for} \quad v \in S\ ,$$

and

$$L(s,\pi,r) = \prod_{\text{all } v} L(s,\pi_v,r_v)$$

(with similar formulas for $L(s,\pi,\tilde{r})$). We note that the factor$L(s,\pi_v,r)$ reduces to the familiar local Langlands factor in case v is *unramified*; indeed in this case, the local coefficient $C_{\bar{\psi}_v}(s,\tilde{\pi}_v)$ is easily computed by way of the formulas already discussed for class 1 Whittaker functions and intertwining operators.

The reader is referred to [Sha8] for a proof of the following fundamental result:

Theorem. The L-function $L(s,\pi,r)$, defined by a convergent Euler product in the half-plane $Re(s) > 2$, extends to a meromorphic function of s in $\mathbb{C}$ with at most *finitely* many poles; moreover, it satisfies the functional equation

(III.2.4.5) $$L(s,\pi,r) = \varepsilon(s,\pi,r)L(1-s,\pi,\tilde{r})$$

with $\varepsilon(s,\pi,r) = \Pi_{v\in S}\varepsilon(s,\pi_v,r,\psi_v)$ as in (III.2.4.4).

The idea of the proof is roughly this:

(1) Rewrite the equation (III.2.1.1) in the form

$$L_S(1+s,\pi,\tilde{r})M(s,\pi)f$$
$$= L_S(s,\pi,\tilde{r}) \bigotimes_{v\in\{\infty\}} \mathcal{A}(s,\pi_v)f_v \left(\bigotimes_{v\in S} A(s,\pi_v)f_v\right) \bigotimes_{v\notin S\cup\{\infty\}} \tilde{f}_v^0$$

where S now contains only finite places, and $\mathcal{A}(s,\pi_v)$ equals the "normalized operator" $\frac{L(1+s,\tilde{r},\varphi_v)}{L(s,\tilde{r},\varphi_v)}A(s,\pi_v)$ for $v \in \{\infty\}$; then the finiteness of poles for $L_S(s,\pi,\tilde{r})$ (in $Re(s) \geq 0$) follows (by an inductive argument) from similar known results for $M(s,\pi)$ together with non-vanishing properties of the normalized operator $\mathcal{A}(s,\pi_v)f_v$ for appropriately chosen f_v;

(2) By a delicate argument (using properties of the local intertwining operators $A(s,\pi_v)$ and local coefficients, $v \in S$) it is shown that multiplication by $\Pi_{v\in S}L(s,\pi_v,r)$ adds at most only finitely many poles to $L_S(s,\pi,r)$;

(3) The functional equation (III.2.4.5) which is obvious from the definitions of ε and $L(s,\pi,r)$, implies (by the monomial nature of $\varepsilon(s,\pi,r)$) that there are also only finitely many poles in the half plane $Re(s) \leq 0$.

The case $m = 2$ is more delicate.

2.5. More examples.

As already indicated before, Theorem 2.4 has many applications to the L-functions heretofore studied, and some new ones as well. We mention a few more here, and refer the reader to [Sha8] for more details.

(i) In [Sha6,7] it is proved (by global methods in [6] and local ones in [7]) that for $G = GL_n \times GL_m \subset GL_{nm}$, the local coefficient $C_\psi(s, \tilde{\pi})$ ($= \gamma_\psi(s, \pi)$) coincides with the local factor $\frac{\varepsilon(s,\pi_1\times\pi_2,\psi)L(s,\pi_1\times\pi_2)}{L(1-s,\tilde{\pi}_1\times\tilde{\pi}_2)}$ of [JPSS2] when $\pi = \pi_1 \otimes \pi_2$. On the one hand, using Theorem 3.1 of [Sha4], this leads to the expected theory for the global Rankin-Selberg convolution $L(s, \pi_1 \times \pi_2)$ initiated in [JPSS2]. On the other hand, it leads to new *local* results concerning intertwining operators and Plancherel measures for $GL(n)$ (cf. [Sha6,7]).

(ii) Let π be a cuspidal representation of $GL_n(\mathbb{A}_F)$ and r either the *second exterior power* (whose restriction to $SL_n(\mathbb{C}) \subset GL_n(\mathbb{C})$has highest weight $\delta_2 = (1, 1, 0 \ldots 0)$) or the *second symmetric power* representation (whose restriction to $SL_n(\mathbb{C})$ has highest weight $2\delta_1 = (2, 0 \ldots 0)$). Then $L(s, \pi, r)$ extends to a meromorphic function which satisfies (III.2.4.5) *and has only finitely many poles in* $\mathbb{C}$. (These are cases (viii) and (iv) of [La1] respectively.)

We note that the L-functions $L(s, \pi, \Lambda^2)$ have also recently been studied using explicit integral expressions ála Rankin-Selberg and Hecke; cf. the unpublished works of [Bump-Friedberg] and [JPSS3]. The latter work gives *some* hope that the (finitely many) poles of $L(s, \pi, \Lambda^2)$ can be explicitly located, thus, too, the poles of $L(s, \pi \times \tau)$ discussed in Section 1.5 above.

(iii) As already described in Section (2.1), Theorem 2.4 applies to the L-functions studied (from a different point of view) in [PSR1] and [PSR2]. However, it seems in general that the method of zeta-integrals might yield more accurate information on the exact location (and nature) of the poles of automorphic L-functions; cf. [PSR2-4] and the discussion in (ii) above. Moreover, the method [PSR1] does not require π to have a Whittaker model.

(iv) At the end of Section II.(2.5) we discussed the symmetric power L-functions for PGL_2, namely $L(s, \pi, Sym^m(\rho_2))$ with $m = 3, 4, 5$. Using the *complete* theory of L-functions ([Sha1-7], plus [JS1, 2] and [GeJa]), it is shown in [Sha9] that $L(s, \pi, Sym^3(\rho_2))$ is actually an *entire* function provided π is "non-monomial" and the Hecke-Jacquet-Langlands L-function $L(s, \pi) \neq 0$ for $s \in (\frac{1}{2}, 1)$. In many cases, this non-vanishing condition is easy to verify, for example for $\pi = \pi_\Delta$, with Δ the Ramanujan cusp form of weight 12 (cf. [Moreno-Shahidi1]). In any case, the relevant diagram data is example (xv) on page 50 of [La1]; cf. our discussion in §II.2.5.

Applications to Shimura-Taniyama-Weil curves (cf. [Shim3]) are also discussed in [Sha9], as are assorted results on the zeros and poles of $L(s, \pi, Sym^4(\rho_2))$ and $L(s, \pi, Sym^5(\rho_2))$ for $Re(s) \geq 1$; cf. [Moreno-Shahidi2,3] and [Rankin2] for classical applications. We do not yet have enough information on the poles of these higher symmetric powers in the half-plane $Re(s) \geq \frac{1}{2}$ to make serious conclusions about their holomorphy in $\mathbb{C}$.

2.6. On uniqueness of the local factors.

In Section 2.4 above, we introduced certain local factors, namely $\gamma(s, \pi_v, v, \bar{\psi}_v)$ and $L(s, \pi_v, r_v)$, in order to "complete" the L-function $L_S(s, \pi, r)$ and describe its "best possible" functional equation (cf. III.2.4.1, 2.4.5, and 2.4.6). In many cases where local factors have alternatively been defined by the method of zeta-integrals, it can be shown that these local factors coincide (cf. 2.4(i) above, for the example $G = GL_n \times GL_m$). A natural question to ask them is, are these local L and ε factors are really canonical at the ramified places? Equivalently, are they uniquely determined by the properties we require of them?

In general, the answer to this question is "yes", but this is not obvious, and we shall attempt only a sketch of the argument below.

Our starting point is the global functional equation (III.2.1.3), together with the definition $\gamma(s,\pi_v,r,\bar{\psi}_v) = C_{\bar{\psi}_v}(s,\bar{\pi}_v)$. For simplicity of exposition, let us assume $m = 1$; then the functional equation reads

$$L_S(s,\pi,r) = \gamma_S(s,\pi,r,\psi)L_S(1-s,\pi,\tilde{r}) \ ,$$

with

$$(*) \qquad \gamma_S(s,\pi,\psi) = \prod_{v\in S} \gamma(s,\pi_v,r,\bar{\psi}_v) \ .$$

Since $L_S(s,\pi,r)$ can not satisfy more than one functional equation of this type, we do know *apriori* that this *product* of local factors $(*)$ is canonical. However, our claim now is that the *local* factors of γ_S are themselves canonically determined. More precisely, we have:

__Claim 1.__ Suppose $\gamma'(s,\pi_v,r,\bar{\psi}_v)$ is defined for every v and such that:

(i) for any archimedean place v, $\gamma'(s,\pi_v,r,\bar{\psi}_v)$ is the appropriate Artin factor coming from the Langlands classification (cf. the beginning of §2.4);

(ii) γ' behaves correctly with respect to induction, in particular, if π_v is induced from the representation $\sigma_1 x \ldots \sigma_\ell$, then $\gamma'(s,\pi_v)$ is factorizable as $\prod_{i=1}^{\ell} \gamma'(s,\sigma_i,r^i)$ for appropriate r^i; and

(iii) whenever π_v is the local component of a global cusp form π, the factor $\gamma'(s,\pi_v,r,\bar{\psi}_v)$ appears in the functional equation for $L_S(s,\pi,r)$. Then

$$\gamma'(s,\pi_v,r,\bar{\psi}_v) = \gamma(s,\pi_v,r,\bar{\psi}_v) \ .$$

__Proof.__ By property (i) we may assume v nonarchimedean, and by property (ii) we may assume π_v supercuspidal. For such a π_v, it can then be shown that π_v is the local component of a *generic* cusp form $\pi = \otimes\pi_v$ with the property that π is unramified at all (finite) places distinct from v. (This is a refinement of Theorem 2.2 of [Vig2], and should appear in a forthcoming work of the second author of this report.) Now apply the functional equation

$$L_S(s,\pi,r) = \left(\prod_{v\in S} \gamma(s,\pi_v,r,\bar{\psi}_v)\right) L_S(1-s,\pi,\tilde{r})$$

with S equal to the union of v with the archimedean places. By properties (i) and (iii) we obtain the desired conclusion.

We turn next to the local Euler factors themselves.

__Claim 2.__ Let $\gamma(s,\pi_v,r,\bar{\psi}_v)$ be as in Claim 1, and assume v nonarchimedean. Let P_v (resp. $\tilde{P}_v$) denote the numerator of the rational function $\gamma(s,\pi_v,r,\bar{\psi}_v)$ (resp. $\gamma(s,\pi_v,\tilde{r},\psi_v)$), normalized so that $P_v(0) = \tilde{P}_v(0) = 1$. Then

$$(*)\qquad \gamma(s,\pi_v,r,\bar{\psi}_v) = \frac{\varepsilon(s,\pi_v,r,\psi_v)L(1-s,\pi_v,\tilde{r})}{L(s,\pi_v,r)}$$

with ε a monomial in q_v^{-s} and *both*

$$L(s,\pi,r) = P_v(q_v^{-s})^{-1}$$

and

$$L(s,\pi,\tilde{r}) = \tilde{P}_v(q_v^{-s})^{-1}$$

independent of ψ_v.

__Proof.__ Recall that the archimedean factors are assumed to be understood in terms of the Langlands classification and the description of local Artin factors;; cf. [Sha4]. By the arguments in Claim 1, we may also assume that π_v is already supercuspidal, and hence a local component of a generic $\pi = \otimes\pi_v$ ramified only at v (and possibly archimedean components). Applying the global functional equation for $L_S(s,\pi,r)$ *twice* then gives the local equation

$$\gamma(s,\pi_v,r,\bar{\psi}_v)\gamma(1-s,\pi,\tilde{r},\psi_v) = 1\ .$$

From this the claim $(*)$ follows, since the zeros of $\tilde{P}_v$ (resp. P_v) must precisely cancel those of the denominator of $\gamma_v(s)$ (resp. $\gamma_v(1-s)$). Similarly, the fact that P_v (and $\tilde{P}_v$) are independent of ψ_v follows from the fact that the zeros of $c_{\psi_v}(s, \pi_v)$ (and hence $\gamma(s, \pi_v, r, \bar{\psi}_v)$) are precisely the poles of the intertwining operator $A(s, \pi_v)$, since π_v is supercuspidal. (Cf. equation II.2.4.1 which relates $c(s, \pi_v)$ and $A(s, \pi_v)$, and immediately implies that the zeros of $c(s, \pi_v)$ are among the poles of $A(s, \pi_v)$. The supercuspidal condition is needed to show that these zeros and poles actually coincide; indeed the results described in §5.4.2 of [Silberger] imply in this case that each pole of $A(s, \pi_v)$ must be a zero of the local coefficient.)

To sum up: Claim 1 asserts that there is no choice in defining local gamma factors for the global functional equation, and Claim 2 says these gamma factors canonically determine the local L-functions. On the other hand, we recall that the theory of zeta-integrals provides a *direct* definition of L-functions as the g.c.d. of a certain family of local zeta-integrals. This definition is independent of the local gamma factors, and yet (by a double application of the local functional equation) one still ends up with a relation of the form

$$\gamma(s, \pi_v, r, \psi_v) = \frac{\varepsilon(s, \pi_v, r, \psi_v) L(1 - s, \pi_v, \tilde{r})}{L(s, \pi_v, r)} .$$

Thus it is natural to ask one last question: Do these two different approaches produce the same local L-functions? For example, when $G = GL_n \times GL_m$, it can be shown that there is no cancellation between the polynomials $L(s, \pi_v, r)^{-1}$ and $L(1 - s, \pi_v, \tilde{r})^{-1}$. It follows that $L(s, \pi_v, r)$ is indeed the (normalized) numerator of $\gamma(s, \pi_v, r, \psi_v)$, and hence that this L-factor must agree with the one defined more generally (for generic representations) in [Sha8]. In general, we expect that $L(s, \pi_v, r)^{-1}$ and $L(1 - s, \pi_v, \tilde{r})^{-1}$ will remain relatively prime, but this remains unproved outside of a few special cases.

Last Words

We have explained how analytic properties of automorphic L-functions may be derived by either the method of constructing explicit zeta-integrals, or by Langlands' method of Eisenstein series. Generally speaking, the advantage of the first method is that it promises to give more precise information on the location of the poles (or even the special values) of these L-functions. On the other hand, Langlands' method has so far applied to a broader class of functions, and seems more suited to harmonic analysis on the group; cf. the papers [Sha6-7] on intertwining operators and Plancherel measures for $GL(n)$. As noted in Sections II.1.8 and 2.6, Langlands' method is also crucial in proving the non-vanishing of automorphic L-functions on the line $Re(s) = 1$; moreover, this method (as opposed to the method of zeta-integrals) is much simpler to apply at the archimedean places (cf. [Sha4]).

Clearly new methods will be required to obtain more precise information on the L-functions already encountered, and to tackle the many L-functions which have thus far resisted the power of existing methods. We mention [PS4] and [PSR4] as interesting new directions but postpone the discussion of these and other works in progress for a future report (see now [Ge5]).

REFERENCES

[Andrianov1] A. Andrianov, "Zeta-functions and the Siegel Modular forms", in *Lie Groups and their Representations*, John Wiley and Sons, 1975.

[Andrianov2] A. Andrianov, "On zeta-functions of Rankin type associated with Siegel Modular forms", in *Lecture Notes in Mathematics*, Springer Verlag, No. 627, 1977.

[Art1] J. Arthur, "Automorphic Representations and Number Theory", in *Canadian Mathematical Society Conference Proceedings*, Vol. 1, Providence, R.I., 1981, pp. 3-51.

[Art2] J. Arthur, "On some problems suggested by the trace formula", in *Lie Group Representations II*, Proceedings University of Maryland, 1982-1983, Lecture Notes in Mathematics, Vol. 1041, Springer Verlag, New York, 1984, pp. 1-49.

[ArtCl] J. Arthur and L. Clozel, "Base Change for $GL(n)$", Preprint, 1986.

[Asai] T. Asai, "On certain Dirichlet series associated with the Hilbert modular forms and Rankin's method, *Math. Ann.* 221 (1977), pp. 81-94.

[BZ1] J. Bernstein and A. Zelevinsky, "Representations of the group $GL(n, F)$ when F is a nonarchimedean local field", *Russian Math. Surveys* 31 (1976), No. 3, pp. 1-68.

[BZ2] J. Bernstein and A. Zelevinsky, "Induced representations of reductive p-adic groups I", *Ann. Sci. ENS* 10 (1977), pp. 441-472.

[Böch] S. Böcherer, "Über die funktionalgleichung automorpher L-funktionen zur Siegelschen modulgruppe", *Journ. für die reine und angewandte Mathematik*, Band 362 (1985), pp. 146-168.

[Bo1] A. Borel, "Automorphic L-functions", in *Proceedings of Symposia in Pure Mathematics*, Vol. 33, Part 2, Amer. Math. Soc., Providence, R.I., 1979, pp. 27-61.

[BoJa] A. Borel and H. Jacquet, "Automorphic forms and automorphic representations", in *Proceedings of Symposia in Pure Math.*, Vol. 33, Part 1, Amer. Math. Soc., Providence, R.I., 1979, pp. 189-202.

[Bump1] D. Bump, "Automorphic Forms on $GL(3, \mathbb{R})$", *Lecture Notes in Math.*, Vol. 1083, Springer Verlag, New York, 1984.

[Bump2] D. Bump, "The Rankin-Selberg Method: A Survey", to appear in *Proceedings of the Symposium in honor of A. Selberg*, Oslo, 1987.

[Bump-Friedberg] D. Bump and S. Friedberg, "The Exterior Square Automorphic L-functions for $GL(n)$", Preprint, 1987.

[Bump-Hoffstein] D. Bump and J. Hoffstein, "Some Euler products associated with cubic metaplectic forms on $GL(3)$", *Duke Math. J.* 53 (1986), p. 1047-1072.

[Car] P. Cartier, "Representations of p-adic groups: A survey", in *Proceedings of Symposia in Pure Mathematics*, Vol. 33, Part 1, Amer. Math. Soc., Providence, R.I., 1979, pp. 111-156.

[CS] W. Casselman and J. Shalika, "The unramified principal series of p-adic groups II: The Whittaker function", *Comp. Math.*, Vol. 41, Fasc. 2 (1980), pp. 207-231.

[Flicker1] Y. Flicker, "Twisted tensors and Euler products", Preprint 1986.

[Flicker2] Y. Flicker, "The adjoint lifting from $SL(2)$ to $PGL(3)$, 1981; also "Symmetric square, I: Definitions and Lemmas, II: Applications of a trace formula, 1984.

[Flicker3] Y. Flicker, *The Trace Formula and Base Change for $GL(3)$*, Springer Lecture Notes in Mathematics, Vol. 927, Springer-Verlag, N.Y. 1982.

[Friedberg] S. Friedberg, "A. global approach to the Rankin-Selberg convolution for $GL(3, \mathbb{Z})$", *Trans. A.M.S.* 300 (1987), pp. 159-178.

[Garrett1] P. Garrett, "Decomposition of Eisenstein series: Rankin triple products", *Annals of Math.* 125 (1987), pp. 209-237.

[Garrett2] P. Garrett, "Integral Representations of Certain L-functions attached to 1, 2 and 3 Modular Forms", Preprint, 1985, Univ. of Minnesota.

[Ge1] S. Gelbart, "An elementary introduction to the Langlands program", *Bulletin (New Series) of the A.M.S.*, Vol. 10, No. 2, April 1984, pp. 177-219.

[Ge2] S. Gelbart, *Automorphic Forms on Adele Groups*, Annals of Mathematics Studies, No. 83, Princeton University Press, 1975.

[Ge3] S. Gelbart, "Examples of dual reductive pairs", *Proc. Symp. in Pure Math.*, Vol.33, Part 1, A.M.S., Providence, 1979, pp. 287-296.

[Ge4] S. Gelbart, "Weil's Representation and the Spectrum of the Metaplectic Group", *Lecture Notes in Mathematics*, Vol. 530, Springer-Verlag, New York, 1976.

[Ge5] S. Gelbart, "Recent Results on Automorphic L-functions", in Proceedings of Symposia in honor of A. Selberg, Oslo, 1987.

[GeJa] S. Gelbart and H. Jacquet, "A relation between automorphic representations of $GL(2)$ and $GL(3)$", *Ann. Sci. Ecole Normale Sup.*, 4e série, Vol. 11 (1978), pp. 471-552.

[GePS1] S. Gelbart and I. Piatetski-Shapiro, "Automorphic forms and L-functions for the Unitary Group", in *Lie Group Representations II, Lecture Notes in Mathematics*, Vol. 1041, Springer-Verlag, New York, 1984, pp. 141-184.

[GePS2] S. Gelbart and I. Piatetski-Shapiro, "On Shimura's correspondence for modular forms of half-integral weight", *Automorphic Forms, Representation Theory and Arithmetic*, (Bombay, 1979), Tata Institute of Fund. Research Studies in Math., No. 10, Bombay, 1981, pp. 1-39.

[GePS3] S. Gelbart and I. Piatetski-Shapiro, "Distinguished representations and modular forms of half-integral weight", *Inventiones Mathematicae*, 59 (1980), pp. 145-188.

[GePS4] S. Gelbart and I. Piatetski-Shapiro, "Some remarks on metaplectic cusp forms and the correspondences of Shimura and Waldspurger", *Israel J. of Math.*, Vol. 44, No. 2, 1983, pp. 97-126.

[GePS5] S. Gelbart and I. Piatetski-Shapiro, "L-functions for $G \times GL(n)$", in *Explicit Constructions of Automorphic L-Functions*, Lecture Notes in Mathematics, Vol. 1254, Springer-Verlag, New York, 1987.

[GelKaz] I.M. Gelfand and D. Kazhdan, "Representations of the group $GL(n, K)$ where K is a local field", in *Lie Groups and their Representations*, Halsted, New York, 1975, pp. 95-118.

[GerLi] P. Gérardin and W. Li, "Fourier transforms of representations of quaternions", *J. für die reine und angewandte Math.*, Band 359 (1985), pp. 121-173.

[God] R. Godement, *Notes on Jacquet-Langlands' Theory*, mimeographed notes, Institute for Advanced Study, 1970.

[GoJa] R. Godement and H. Jacquet, "Zeta-functions of Simple Algebras", *Lecture Notes in Mathematics*, Vol. 260, Springer-Verlag, New York, 1972.

[HLR] G. Harder, R.P. Langlands and M. Rapoport, "Algebraische zykeln auf Hilbert-Blumenthal-Flaschen", *J. Reine Angew. Math.* 366 (1986), pp. 53-120.

[Harish-Chandra] Harish-Chandra, "Harmonic Analysis on Reductive p-adic Groups", *Proc. Sympos. Pure Math.*, vol. 26, Amer. Math. Soc., Providence, R.I., 1973, pp. 167-192.

[Henniart] G. Henniart, "On the local Langlands conjecture for $GL(n)$: the cyclic case", *Annals of Mathematics*, 123 (1986), pp. 145-203.

[Howe] R. Howe, "θ-series and invariant theory", *Proc. of Symp. in Pure Math.*, Vol. 33, Part 1, A.M.S., Providence, 1979, pp. 275-286.

[Jacquet] H. Jacquet, "Automorphic Forms on $GL(2)$, II", *Lecture Notes in Mathematics*, Vol. 278, Springer-Verlag, New York, 1972.

[JL] H. Jacquet and R.P. Langlands, "Automorphic forms on $GL(2)$", *Lecture Notes in Mathematics*, Vol. 114, Springer-Verlag, New York, 1970.

[JPSS1] H. Jacquet, I. Piatetski-Shapiro and J. Shalika, "Automorphic Forms on $GL(3)$ II", *Annals of Math.* 109 (1979), pp. 213-258.

[JPSS2] H. Jacquet, I. Piatetski-Shapiro and J. Shalika, "Rankin-Selberg Convolutions", *Amer. J. Math.*, Vol. 105 (1983), pp. 367-464.

[JPSS3] H. Jacquet, I. Piatetski-Shapiro and J. Shalika, work in progress on lifting symplectic forms to $GL(4)$.

[JPSS4] H. Jacquet, I. Piatetski-Shapiro and J. Shalika, "Facteurs L and ε du groupe linéaire: théorie archimédienne", *C.R.A.S.* (série A) 293 (1981), pp. 13-18.

[JPSS5] H. Jacquet, I. Piatetski-Shapiro and J. Shalika, "Relevement cubique nonnormal", *C.R. Acad. Sci. Paris*, Sér. I. Math. 292 (1981), No. 12, pp. 567-571.

[JS1] H. Jacquet and J. Shalika, "On Euler Products and the Classification of Automorphic Representations I", *Amer. J. Math.*, Vol. 103, No. 3, (1981), pp. 499-558.

[JS2] H. Jacquet and J. Shalika, "On Euler Products and the Classification of Automorphic Representations II", *Amer. J. Math.*, Vol. 103, No. 3, (1981), pp. 777-815.

[JS3] H. Jacquet and J. Shalika, "A non-vanishing theorem for zeta-functions of GL_n, *Inventiones Math.*, Vol. 38 (1976), pp. 1-16.

[Ka] S. Kato, "On an explicit formula for class-1 Whittaker functions on split reductive groups over p-adic fields", preprint, 1978, University of Tokyo.

[Katz] N. Katz, Review of P. Deligne's "La Conjecture de Weil, I", *Math. Reviews*, Vol. 49, #5013, 1975.

[Keys-Sha] D. Keys and F. Shahidi, "Artin L-functions and Normalization of Intertwining Operators", preprint 1986.

[Kubota] T. Kubota, *Automorphic Forms and Reciprocity in a Number Field*, Kiyokuniya Book Store and Kyoto University, 1969.

[Kud1] S. Kudla, "On certain Euler products for $SU(2,1)$", *Comp. Math.* 42 (1981), pp. 321-344.

[Kud2] S. Kudla, "Periods of Integrals for $SU(n,1)$", *Comp. Math.* 50 (1983), pp. 3-63.

[LabesseLa] J.P. Labesse and R.P. Langlands, "L-indistinguishability for $SL(2)$", *Canad. J. Math.* 31 (1979), pp. 726-785.

[Lai] K.F. Lai, 11Tamagawa number of reductive algebraic groups", *Comp. Math.* 41 (1980), pp. 153-188.

[La1] R.P. Langlands, *Euler Products*, Yale University Press, James K. Whitmore Lectures, 1967.

[La2] R.P. Langlands, "On the functional equations satisfied by Eisenstein series", Springer Lecture Notes in Mathematics, Vol. 544, 1976.

[La3] R.P. Langlands, "Problems in the Theory of Automorphic Forms", in *Lecture Notes in Mathematics*, Vol. 170, Springer-Verlag, New York, 1970, pp. 18-86.

[La4] R.P. Langlands, "L-functions and automorphic representations", *Proc. Int. Congress Math.*, Helsinki, 1978.

[La5] R.P. Langlands, "Automorphic representations, Shimura varieties and motives", *Proc. Symp. Pure Math.*, Vol. 33, Part 2, Amer. Math. Soc., Providence, R.I., 1979, pp. 205-246.

[La6] R.P. Langlands, "On the classification of irreducible representations of real reductive groups", mimeographed notes, Institute for Advanced Study, 1973.

[La7] R.P. Langlands, "On Artin's L-functions", Rice University Studies 56 (1970), pp. 23-28.

[La8] R.P. Langlands, On the notion of an automorphic representation", *Proc. Symp. Pure Math.*, Vol. 33, Part 1, A.M.S., Providence, R.I., 1979, pp. 203-208.

[La9] R.P. Langlands, *Base Change for GL(2)*, Annals of Math. Studies, No. 96, Princeton Univ. Press, 1980.

[La10] R.P. Langlands, Les débuts d'une formule des traces stables", Publ. Math. de l'Université Paris VII, No. 13 (1983).

[Li1] W. Li, "On the representations of $GL(2)$, I and II", *J. für die reine und angewandte Mathematik*, Band 313 (1980), pp. 27-42, and Band 314 (1980), pp. 3-20.

[Li2] W. Li, "On converse theorems for $GL(2)$ and $GL(1)$", *Amer. J. Math.* 163, No. 5 (1981), pp. 851-885.

[Lion-Vergne] G. Lion and M. Vergne, *The Weil representation, Maslov index and theta-series*, Progress in Math., Vol. 6, Birkhäuser, Boston, 1980.

[Moeg-Wald] C. Moeglin et J.L. Waldspurger, "Modeles de Whittaker degeneres pour des groupes p-adiques", preprint, 1986, Paris.

[MVW] C. Moeglin, M.-F. Vignéras and J.-L. Waldspurger, "Correspondences de Howe sur un corps p-adique", *Lecture Notes in Mathematics*, Vol. 1291, Springer-Verlag, New York, 1987.

[Moreno1] C. Moreno, "An analytic proof of the Strong Multiplicity One Theorem", *Amer. J. Math.*, Vol. 107 (1985), pp. 163-206.

[Moreno2] C. Moreno, "Meromorphic Continuation and Functional Equation of certain automorphic L-functions", preliminary report, 1983.

[Moreno-Shahidi1] C.J. Moreno and F. Shahidi, "The L-function $L_3(s, \pi_\Delta)$ is entire", *Inventiones Math.*, 79 (1985), pp. 247-251.

[Moreno-Shahidi2] C.J. Moreno and F. Shahidi, "The L-functions $L(s, Sym^m(r), \pi)$", *Canadian Math. Bull.* 28 (1985), pp. 405-410.

[Moreno-Shahidi3] C.J. Moreno and F. Shahidi, "The fourth moment of Ramanujan's τ-function", Math. Ann. 266 (1983), pp. 233-239.

[Niwa] S. Niwa, "Modular forms of half-integral weight and the integral of certain theta-functions", *Nagoya Math. J.* 56 (1975), pp. 147-161.

[Novodvorsky] M. Novodvorsky, Automorphic L-functions for the symplectic group GSp_4, in *Proc. Symp. in Pure Math.*, Vol. 33, Part 2, 1979, A.M.S., Providence, R.I., pp. 87-96.

[PattPS1] S.J. Patterson and I. Piatetski-Shapiro, "A cubic analog of the cuspidal theta-series", *J. Math. Purc et Appl.* 63 (1984), pp. 333-375.

[PattPS2] S.J. Patterson and I. Piatetski-Shapiro, "Symmetric square L-functions for $GL(3)$", preprint, 1987.

[PS1] I. Piatetski-Shapiro, "Converse theorem for $GL(3)$", Lecture Note. #15, University of Maryland, Department of Mathematics, 1975.

[PS2] I. Piatetski-Shapiro, "Euler Subgroups", in *Lie Groups and their Representations*, Halsted, New York, 1975.

[PS3] I. Piatetski-Shapiro, "Tate Theory for Reductive Groups and Distinguished Representations", *Proceedings of the International Congress of Math.*, Helsinki, 1978.

[PS4] I. Piatetski-Shapiro, "Invariant Theory and Kloosterman Sums", Preprint, 1987.

[PS5] I. Piatetski-Shapiro, "Multiplicity One Theorems", in *Proceedings of Symp. in Pure Math.*, Vol. 33 (1979), Part 1, pp. 209-212, A.M.S., Providence, R.I.

[PSR1] I. Piatetski-Shapiro and S. Rallis, *L-functions for the Classical Groups*, notes prepared by J. Cogdell, in "Explicit Constructions of Automorphic L-functions", *Lecture Notes in Mathematics*, Vol. 1254, Springer-Verlag, 1987.

[PSR2] I. Piatetski-Shapiro and S. Rallis, "ε-factors of Representations of Classical Groups", *Proceedings of the National Academy of Sciences*, U.S.A., 83 (1986), pp. 4589-4593.

[PSR3] I. Piatetski-Shapiro and S. Rallis, "Rankin Triple L-functions, *Comp. Math.* 64 (1987), pp. 31-115.

[PSR4] I. Piatetski-Shapiro and S. Rallis, "A new way to get Euler Products, preprint, Fall 1986.

[PSS1] I. Piatetski-Shapiro and D. Soudry, "L and ε functions for $GSp(4) \times GL(2)$", *Proc. Nat. Acad. Sci.*, U.S.A., Vol. 82, June 1984, pp. 3924-3927.

[PSS2] I. Piatetski-Shapiro and D. Soudry, "Automorphic Forms on the Symplectic Groups of Order four", preprint, Institut des Hautes Etudes Scientifiques, Bures-sur-Yvette, July 1983.

[Rallis1] S. Rallis, "Langlands' functoriality and the Weil representation", *Amer. J. Math.*, 104 (1982), pp. 469-515.

[Rallis2] S. Rallis, "On the Howe duality conjecture", *Compositio Math.* 51 (1984), pp. 333-399; also "Injectivity properties of liftings associated to Weil representations", *Compositio Math.* 52 (1984), pp. 139-169.

[Rallis3] S. Rallis, "*L*-functions and the Oscillator Representations", *Lecture Notes in Mathematics*, Vol. 1245, Springer-Verlag, 1987.

[Rankin1] R. Rankin, "Contributions to the theory of Ramanujan's function $\tau(n)$ and similar arithmetical functions, I and II", Proc. Camb. Phil. Soc. 35 (1939), pp. 351-356 and 357-372.

[Rankin2] R. Rankin, "Fourier coefficients of cusp forms", in *Math. Proc. Camb. Phil. Soc.* (1986), Vol. 100, Part 1, pp. 5-29.

[RapZink] M. Rapoport and Th. Zink, "Über die lokale zetafunktion....", *Inv. Mathematicae* 68 (1982), pp. 21-101.

[Rodier1] F. Rodier, "Decomposition of principal series for reductive *p*-adic groups, and the Langlands' classification", in: *Operator Algebras and Group Representations*, Vol. II (Neptun, 1980), Monographs Stud.Math., 18, Pitman, Boston, Mass.-London, 1984, pp. 86-94.

[Rodier2] F. Rodier, "Modéle de Whittaker et caractéres de représentations" in *Lecture Notes in Mathematics*, Vol. 466, Springer Verlag, 1974, pp. 151-171.

[Rogawski] J. Rogawski, "Automorphic Representations of Unitary Groups in three variables", *Annals of Mathematics Studies*, to appear.

[Satake] I. Satake, "Theory of spherical functions on reductive algebraic groups over *p*-adic fields", *Publ. Math. IHES* 18 (1963), pp. 1-69.

[Selberg] A. Selberg, "Bemerkungen über eine Dirichletsche Reihe, die mit der Theorie der Modulformen nahe verbunden ist", Arch. Math. Naturvid. 43 (1940), pp. 47-50.

[Serre] J.P. Serre, *Abelian ℓ-adic Representations*, W.A. Benjamin, Inc., New York, Amsterdam, 1968.

[Sha1] F. Shahidi, "Functional Equation satisfied by certain L-functions", *Comp. Math.* 37 (1978), pp. 171-208.

[Sha2] F. Shahidi, "Whittaker models for real groups", *Duke Math. J.*, 47 (1980), pp. 99-125.

[Sha3] F. Shahidi, "On Certain L-functions", *Amer. J. Math.*, Vol. 103, No. 2, 1981, pp. 297-355.

[Sha4] F. Shahidi, "Local coefficients as Artin factors for real groups", *Duke Math. J.* 52 (1985), pp. 973-1007.

[Sha5] F. Shahidi, "Artin L-functions and normalization of intertwining operators", in "Seminar on the Analytical Aspects of the Trace Formula II", Institute for Advanced Study, 1983-84.

[Sha6] F. Shahidi, "Local coefficients and normalization of intertwining operators for $GL(n)$", *Comp. Math.* 48 (1983), pp. 271-295.

[Sha7] F. Shahidi, "Fourier transforms of intertwining operators and Plancherel measures for $GL(n)$", *Amer. J. Math.* 106 (1984), pp. 67-111.

[Sha8] F. Shahidi, "On the Ramanujan Conjecture and Finiteness of Poles for Certain L-functions", *Annals of Math.*, to appear.

[Sha9] F. Shahidi, Third symmetric power L-functions for $GL(2)$", to appear.

[Shalika] J. Shalika, "The Multiplicity One theorem for $GL(n)$", *Annals of Math.* 100 (1974), pp. 171-193.

[Shim1] G. Shimura, "On modular forms of half-integral weight", *Annals of Math.*, Vol. 97 (1973), pp. 440-481.

[Shim2] G. Shimura, "On the holomorphy of certain Dirichlet series", *Proc. London Math. Soc.* (3) (1975), pp. 79-98.

[Shim3] G. Shimura, "On the factors of the Jacobian variety of a modular function field", *J. Math. Soc. Japan* 25 (1973), pp. 523-544.

[Shin1] T. Shintani, "On an explicit formula for class 1 Whittaker functions on GL_n over p-adic fields", *Proc. Jap. Acad.* 52 (1976), pp. 180-182.

[Shin2] T. Shintani, "On construction of holomorphic cusp forms of half-integral weight", *Nagoya Math. J.* 58 (1975), pp. 83-126.

[Silberger] A. Silberger, *Introduction to harmonic analysis on reductive p-adic groups.* Math. Notes of Princeton University Press 23, Princeton, 1979.

[Soudry] D. Soudry, "The L and γ factors for generic representations of $GSp(4,k) \times GL(2,k)$ over a local nonarchimedean field k", *Duke Math. J.*, Vol. 51, No. 2, June 1984, pp. 355-394.

[Spr1] T. Springer, *Linear Algebraic Groups*, Birkhauser, Boston, 1982.

[Spr2] T. Springer, "Reductive Groups", in *Proc. Symp. in Pure Math.*, Vol. 33, Part 1, A.M.S., Providence, R.I., 1979, pp. 3-27.

[Tamagawa] T. Támagawa, "The zeta-function of a simple algebra", *Annals of Math.* 77 (1963), pp. 387-405.

[Tate1] J. Tate, "Fourier analysis in number fields and Hecke's zeta-function", Thesis, Princeton, 1950, also appears in *Algebraic Number Theory*, Proceedings of the Brighton Conference, Edited by J.W. Cassels and A. Frohlich, Academic Press, New York, 1968.

[Tate2] J. Tate, "Number Theoretic Background", in *Proc. Symp. in Pure Math.*, Vol. 33, Part 2, Amer. Math. Soc., Providence, R.I., 1979, pp. 3-26.

[Tunnell] J. Tunnell, "Artin's conjecture for representations of octahedral type", *Bull. Amer. Math. Soc.* 5 (1981), pp. 173-175.

[Vig1] M.-F. Vignéras, "Quelques remarques sur la conjecture $\lambda_1 \geq \frac{1}{4}$", in *Sém. Théorie des Nombres*, Paris 1981-82, M.-J. Bertin (Ed.), Birkhäuser, Boston, 1983, pp. 321-343.

[Vig2] M.-F. Vignéras, "Correspondences entres representations automorphes de $GL(2)$ sur une extension quadratique et de $GSp(4)$ sur D; conjecture locale de Langlands pour $GSp(4)$", *The Selberg Trace Formula and Related Topics*, Contemporary Mathematics, Vol. 53, A.M.S., Providence, R.I., 1986.

[Wald] J.L. Waldspurger, "Correspondance de Shimura", *J. Math. Pures Appl.* 59 (1980), pp. 1-133.

[We1] A. Weil, "Über die Bestimmuag Dirichletscher Reihen durch Funktionalgleichungen", *Math. Ann.* 168(1967), pp. 149-156.

[We2] A. Weil, "Sur certain groupes d'opérateurs unitaries", *Acta Math.* 111 (1964), pp. 143-2.

[We3] A. Weil, "Sur la formule de Siegel dans le theorie des groupes classiques", *Acta Math.* 113 (1965), pp. 1-87.

INDEX

Perspectives in Mathematics

Vol. 1 J. S. Milne, *Arithmetic Duality Theory*
Vol. 2 A. Borel et al., *Algebraic D-Modules*
Vol. 3 Ehud de Shalit, *Iwasawa Theory of Elliptic Curves with Complex Multiplication*
Vol. 4 M. Rapoport, N. Schappacher, P. Schneider, editors, *Beilinson's Conjectures on Special Values of L-Functions*
Vol. 5 Paul Günther, *Huygens' Principle and Hyperbolic Equations*
Vol. 6 Stephen Gelbart and Freydoon Shahidi, *Analytic Properties of Automorphic L-Functions*